Computer-Integrated Manufacturing

Computer-Integrated Manufacturing

Henry Webber

NY RESEARCH PRESS
New York

Published by NY Research Press
118-35 Queens Blvd., Suite 400,
Forest Hills, NY 11375, USA
www.nyresearchpress.com

Computer-Integrated Manufacturing
Henry Webber

International Standard Book Number: 978-1-63238-817-9 (Hardback)

Cataloging-in-Publication Data

Computer-integrated manufacturing / Henry Webber.
p. cm.
Includes bibliographical references and index.
ISBN 978-1-63238-817-9
1. Computer integrated manufacturing systems. 2. Computer-aided engineering.
3. Industrial engineering. I. Webber, Henry.
TS155.63 .C66 2020
670.285--dc23

Printed in China.

Contents

Preface

The approach towards manufacturing which aims to automate the entire production process using computers is known as computer-integrated manufacturing. This system uses computers to link materials handling and management with functional areas such as inventory control, distribution, planning, purchasing, design and analysis. It finds application in a variety of industries like ship building, automotive and aviation industries. There are three key challenges which are faced in the development of a smoothly operating computer integrated manufacturing system. These are integrating components from different suppliers, data integrity and process control. A computer-integrated manufacturing system makes use of a number of subsystems such as computer aided design, enterprise resource planning, computer-aided quality assurance and computer-aided engineering. This book unfolds the innovative aspects of computer-integrated manufacturing which will be crucial for the progress of this field in the future. It elucidates new techniques and their applications in a multidisciplinary approach. Coherent flow of topics, student-friendly language and extensive use of examples make this book an invaluable source of knowledge.

A detailed account of the significant topics covered in this book is provided below:

Chapter 1- Automation is the technology that uses automatic control system for operating equipment like machinery, processes in factories, telephone networks and stabilization of ships and aircrafts. It consists of three major elements- power, program of instructions and a control system. This is an introductory chapter which will briefly introduce about automation and its elements.

Chapter 2- Computer-Integrated manufacturing deals with the application of computer-controlled machinery and software for automation in manufacturing processes. A few of its types are computer-aided design, computer-aided manufacturing, computer-aided process planning, etc. This chapter discusses the varied types of computer-integrated manufacturing in detail.

Chapter 3- Industrial control systems deal with the operation of control systems and instrumentation used in industrial processes. It includes discrete process control, computer process control, computer numerical control, motion control system, supervisory control, artificial neural network, etc. The topics elaborated in this chapter will help in gaining a better perspective of these industrial control systems.

Chapter 4- Automation in production systems is concerned with operations and control of mechanical, industrial, electronic and computer based systems. Industrial automation uses information technologies and computers for handling operations in an industry. This chapter has been carefully written to provide an easy understanding of automation in production systems.

Chapter 5- There are wide-range of applications of automation in the fields of retail, mining, highway systems, car and home systems, waste management, power production, etc. This chapter closely examines the diverse applications of automation to provide an extensive understanding of the subject.

I would like to make a special mention of my publisher who considered me worthy of this opportunity and also supported me throughout the process. I would also like to thank the editing team at the back-end who extended their help whenever required.

Henry Webber

1

Automation: An Introduction

Automation is the technology that uses automatic control system for operating equipment like machinery, processes in factories, telephone networks and stabilization of ships and aircrafts. It consists of three major elements- power, program of instructions and a control system. This is an introductory chapter which will briefly introduce about automation and its elements.

Automation is the application of machines to tasks once performed by human beings or, increasingly, to tasks that would otherwise be impossible. Although the term mechanization is often used to refer to the simple replacement of human labour by machines, automation generally implies the integration of machines into a self-governing system. Automation has revolutionized those areas in which it has been introduced, and there is scarcely an aspect of modern life that has been unaffected by it.

The term automation was coined in the automobile industry about 1946 to describe the increased use of automatic devices and controls in mechanized production lines. The origin of the word is attributed to D.S. Harder, an engineering manager at the Ford Motor Company at the time. The term is used widely in a manufacturing context, but it is also applied outside manufacturing in connection with a variety of systems in which there is a significant substitution of mechanical, electrical, or computerized action for human effort and intelligence.

In general usage, automation can be defined as a technology concerned with performing a process by means of programmed commands combined with automatic feedback control to ensure proper execution of the instructions. The resulting system is capable of operating without human intervention. The development of this technology has become increasingly dependent on the use of computers and computer-related technologies. Consequently, automated systems have become increasingly sophisticated and complex. Advanced systems represent a level of capability and performance that surpass in many ways the abilities of humans to accomplish the same activities.

Automation technology has matured to a point where a number of other technologies have developed from it and have achieved a recognition and status of their own. Robotics is one of these technologies; it is a specialized branch of automation in which the automated machine possesses certain anthropomorphic, or humanlike, characteristics. The most typical humanlike characteristic of a modern industrial robot is its powered mechanical arm. The robot's arm can be programmed to move through a sequence of motions to perform useful tasks, such as loading and unloading parts at a production machine or making a sequence of spot-welds on the sheet-metal parts of an automobile body during assembly. As these examples suggest, industrial robots are typically used to replace human workers in factory operations.

Principles and Theory of Automation

The developments described above have provided the three basic building blocks of automation: (1) a source of power to perform some action, (2) feedback controls, and (3) machine programming. Almost without exception, an automated system will exhibit all these elements.

Power Source

An automated system is designed to accomplish some useful action, and that action requires power. There are many sources of power available, but the most commonly used power in today's automated systems is electricity. Electrical power is the most versatile, because it can be readily generated from other sources (e.g., fossil fuel, hydroelectric, solar, and nuclear) and it can be readily converted into other types of power (e.g., mechanical, hydraulic, and pneumatic) to perform useful work. In addition, electrical energy can be stored in high-performance, long-life batteries.

The actions performed by automated systems are generally of two types: (1) processing and (2) transfer and positioning. In the first case, energy is applied to accomplish some processing operation on some entity. The process may involve the shaping of metal, the molding of plastic, the switching of electrical signals in a communication system, or the processing of data in a computerized information system. All these actions entail the use of energy to transform the entity (e.g., the metal, plastic, electrical signals, or data) from one state or condition into another more valuable state or condition. The second type of action—transfer and positioning—is most readily seen in automated manufacturing systems designed to perform work on a product. In these cases, the product must generally be moved (transferred) from one location to another during the series of processing steps. At each processing location, accurate positioning of the product is generally required. In automated communications and information systems, the terms transfer and positioning refer to the movement of data (or electrical signals) among various processing units and the delivery of information to output terminals (printers, video display units, etc.) for interpretation and use by humans.

Feedback Controls

Feedback controls are widely used in modern automated systems. A feedback control system consists of five basic components: (1) input, (2) process being controlled, (3) output, (4) sensing elements, and (5) controller and actuating devices. These five components are illustrated in figure. The term closed-loop feedback control is often used to describe this kind of system.

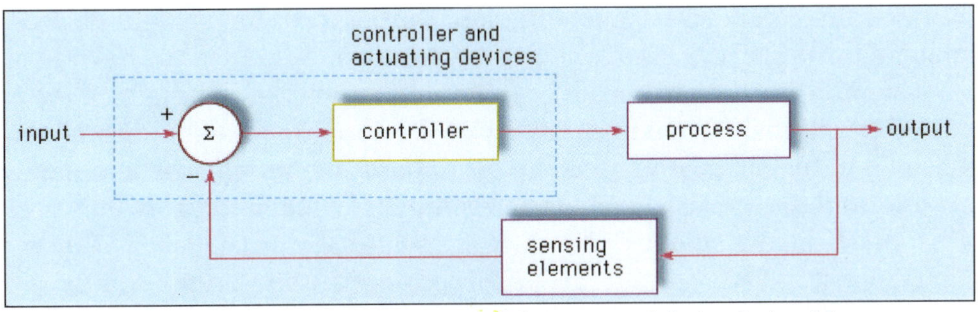

The components of a feedback control system and their relationships.

The input to the system is the reference value, or set point, for the system output. This represents the desired operating value of the output. Using the previous example of the heating system as an illustration, the input is the desired temperature setting for a room. The process being controlled is the heater (e.g., furnace). In other feedback systems, the process might be a manufacturing operation, the rocket engines on a space shuttle, the automobile engine in cruise control, or any of a variety of other processes to which power is applied. The output is the variable of the process that is being measured and compared to the input; in the above example, it is room temperature.

The sensing elements are the measuring devices used in the feedback loop to monitor the value of the output variable. In the heating system example, this function is normally accomplished using a bimetallic strip. This device consists of two metal strips joined along their lengths. The two metals possess different thermal expansion coefficients; thus, when the temperature of the strip is raised, it flexes in direct proportion to the temperature change. As such, the bimetallic strip is capable of measuring temperature. There are many different kinds of sensors used in feedback control systems for automation.

The purpose of the controller and actuating devices in the feedback system is to compare the measured output value with the reference input value and to reduce the difference between them. In general, the controller and actuator of the system are the mechanisms by which changes in the process are accomplished to influence the output variable. These mechanisms are usually designed specifically for the system and consist of devices such as motors, valves, solenoid switches, piston cylinders, gears, power screws, pulley systems, chain drives, and other mechanical and electrical components. The switch connected to the bimetallic strip of the thermostat is the controller and actuating device for the heating system. When the output (room temperature) is below the set point, the switch turns on the heater. When the temperature exceeds the set point, the heat is turned off.

Machine Programming

The programmed instructions determine the set of actions that is to be accomplished automatically by the system. The program specifies what the automated system should do and how its various components must function in order to accomplish the desired result. The content of the program varies considerably from one system to the next. In relatively simple systems, the program consists of a limited number of well-defined actions that are performed continuously and repeatedly in the proper sequence with no deviation from one cycle to the next. In more complex systems, the number of commands could be quite large, and the level of detail in each command could be significantly greater. In relatively sophisticated systems, the program provides for the sequence of actions to be altered in response to variations in raw materials or other operating conditions.

Programming commands are related to feedback control in an automated system in that the program establishes the sequence of values for the inputs (set points) of the various feedback control loops that make up the automated system. A given programming command may specify the set point for the feedback loop, which in turn controls some action that the system is to accomplish. In effect, the purpose of the feedback loop is to verify that the programmed step has been carried out. For example, in a robot controller, the program might specify that the arm is to move to a designated position, and the feedback control system is used to verify that the move has been correctly made. The relationship of program control and feedback control in an automated system is illustrated in figure.

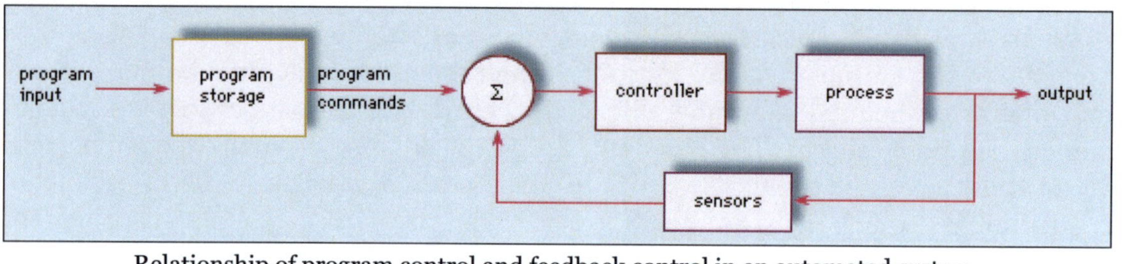

Relationship of program control and feedback control in an automated system.

Some of the programmed commands may be executed in a simple open-loop fashion—i.e., without the need for a feedback loop to verify that the command has been properly carried out. For example, a command to flip an electrical switch may not require feedback. The need for feedback control in an automated system might arise when there are variations in the raw materials being fed into a production process, and the system must take these variations into consideration by making adjustments in its controlled actions. Without feedback, the system would be unable to exert sufficient control over the quality of the process output.

The programmed commands may be contained on mechanical devices (e.g., mechanical cams and linkages), punched paper tape, magnetic tape, magnetic disks, computer memory, or any of a variety of other media that have been developed over the years for particular applications. It is common today for automated equipment to use computer storage technology as the means for storing the programmed commands and converting them into controlled actions. One of the advantages of computer storage is that the program can be readily changed or improved. Altering a program that is contained on mechanical cams involves considerable work.

Programmable machines are often capable of making decisions during their operation. The decision-making capacity is contained in the control program in the form of logical instructions that govern the operation of such a system under varying circumstances. Under one set of circumstances, the system responds one way; under different circumstances, it responds in another way. There are several reasons for providing an automated system with decision-making capability, including (1) error detection and recovery, (2) safety monitoring, (3) interaction with humans, and (4) process optimization.

Error detection and recovery is concerned with decisions that must be made by the system in response to undesirable operating conditions. In the operation of any automated system, malfunctions and errors sometimes occur during the normal cycle of operations, for which some form of corrective action must be taken to restore the system. The usual response to a system malfunction has been to call for human assistance. There is a growing trend in automation and robotics to enable the system itself to sense these malfunctions and to correct for them in some manner without human intervention. This sensing and correction is referred to as error detection and recovery, and it requires that a decision-making capability be programmed into the system.

Safety monitoring is a special case of error detection and recovery in which the malfunction involves a safety hazard. Decisions are required when the automated system sensors detect that a safety condition has developed that would be hazardous to the equipment or humans in the vicinity of the equipment. The purpose of the safety-monitoring system is to detect the hazard and to take the most appropriate action to remove or reduce it. This may involve stopping the operation and alerting maintenance personnel to the condition, or it may involve a more complex set of actions to eliminate the safety problem.

Automated systems are usually required to interact with humans in some way. An automatic bank teller machine, for example, must receive instructions from customers and act accordingly. In some automated systems, a variety of different instructions from humans is possible, and the decision-making capability of the system must be quite sophisticated in order to deal with the array of possibilities.

A fourth reason for decision making in an automated system is to optimize the process. The need for optimization occurs most commonly in processes in which there is an economic performance criterion whose optimization is desirable. For example, minimizing cost is usually an important objective in manufacturing. The automated system might use adaptive control to receive appropriate sensor signals and other inputs and make decisions to drive the process toward the optimal state.

Industrial Robotics

Industrial robotics is an automation technology that has received considerable attention since about 1960.

Development of Robotics

Robotics is based on two related technologies: numerical control and teleoperators. Numerical control (NC) is a method of controlling machine tool axes by means of numbers that have been coded on punched paper tape or other media. It was developed during the late 1940s and early 1950s. The first numerical control machine tool was demonstrated in 1952 in the United States at the Massachusetts Institute of Technology (MIT). Subsequent research at MIT led to the development of the APT (Automatically Programmed Tools) language for programming machine tools.

A teleoperator is a mechanical manipulator that is controlled by a human from a remote location. Initial work on the design of teleoperators can be traced to the handling of radioactive materials in the early 1940s. In a typical implementation, a human moves a mechanical arm and hand at one location, and these motions are duplicated by the manipulator at another location.

Industrial robotics can be considered a combination of numerical-control and teleoperator technologies. Numerical control provides the concept of a programmable industrial machine, and teleoperator technology contributes the notion of a mechanical arm to perform useful work. The first industrial robot was installed in 1961 to unload parts from a die-casting operation. Its development was due largely to the efforts of the Americans George C. Devol, an inventor, and Joseph F. Engelberger, a businessman. Devol originated the design for a programmable manipulator, the U.S. patent for which was issued in 1961. Engelberger teamed with Devol to promote the use of robots in industry and to establish the first corporation in robotics—Unimation, Inc.

The Robot Manipulator

The most widely accepted definition of an industrial robot is one developed by the Robotic Industries Association:

> "An industrial robot is a reprogrammable, multifunctional manipulator designed to move materials, parts, tools, or specialized devices through variable programmed motions for the performance of a variety of tasks".

The technology of robotics is concerned with the design of the mechanical manipulator and the computer systems used to control it. It is also concerned with the industrial applications of robots.

The mechanical manipulator of an industrial robot is made up of a sequence of link and joint combinations. The links are the rigid members connecting the joints. The joints (also called axes) are the movable components of the robot that cause relative motion between adjacent links. There are five principal types of mechanical joints used to construct the manipulator. Two of the joints are linear, in which the relative motion between adjacent links is translational, and three are rotary types, in which the relative motion involves rotation between links.

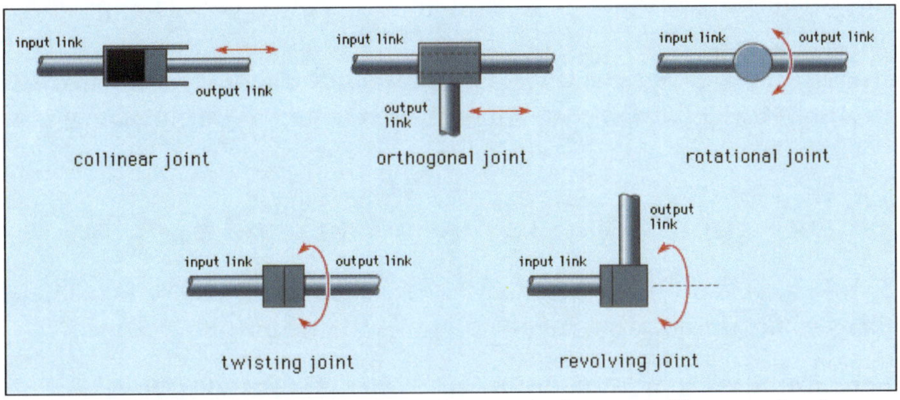

Five types of mechanical joints used in robot manipulators. Collinear and orthogonal are translational joints; rotational, twisting, and revolving are rotary joints.

The manipulator can be divided into two sections: (1) an arm-and-body, which usually consists of three joints connected by large links, and (2) a wrist, consisting of two or three compact joints. Attached to the wrist is a gripper to grasp a work part or a tool (e.g., a spot-welding gun) to perform a process. The two manipulator sections have different functions: the arm-and-body is used to move and position parts or tools in the robot's work space, while the wrist is used to orient the parts or tools at the work location. The arm-and-body section of most commercial robots is based on one of four configurations. Each of the anatomies, as they are sometimes called, provides a different work envelope (i.e., the space that can be reached by the robot's arm) and is suited to different types of applications.

Robot Programming

The computer system that controls the manipulator must be programmed to teach the robot the particular motion sequence and other actions that must be performed in order to accomplish its task. There are several ways that industrial robots are programmed. One method is called lead-through programming. This requires that the manipulator be driven through the various motions needed to perform a given task, recording the motions into the robot's computer memory. This can be done either by physically moving the manipulator through the motion sequence or by using a control box to drive the manipulator through the sequence.

A second method of programming involves the use of a programming language very much like a computer programming language. However, in addition to many of the capabilities of a computer programming language (i.e., data processing, computations, communicating with other computer devices, and decision making), the robot language also includes statements specifically designed for robot control. These capabilities include (1) motion control and (2) input/output. Motion-control commands are used to direct the robot to move its manipulator to some defined position in space. For example, the statement "move P1" might be used to direct the robot to a point in space called P1. Input/output commands are employed to control the receipt of signals from sensors and other devices in the work cell and to initiate control signals to other pieces of equipment in the cell. For instance, the statement "signal 3, on" might be used to turn on a motor in the cell, where the motor is connected to output line 3 in the robot's controller.

Manufacturing Applications of Automation and Robotics

One of the most important application areas for automation technology is manufacturing. To many people, automation means manufacturing automation.

Three types of automation in production can be distinguished: (1) fixed automation, (2) programmable automation, and (3) flexible automation.

Fixed automation, also known as "hard automation," refers to an automated production facility in which the sequence of processing operations is fixed by the equipment configuration. In effect, the programmed commands are contained in the machines in the form of cams, gears, wiring, and other hardware that is not easily changed over from one product style to another. This form of automation is characterized by high initial investment and high production rates. It is therefore suitable for products that are made in large volumes. Examples of fixed automation include machining transfer lines found in the automotive industry, automatic assembly machines, and certain chemical processes.

Programmable automation is a form of automation for producing products in batches. The products are made in batch quantities ranging from several dozen to several thousand units at a time. For each new batch, the production equipment must be reprogrammed and changed over to accommodate the new product style. This reprogramming and changeover take time to accomplish, and there is a period of nonproductive time followed by a production run for each new batch. Production rates in programmable automation are generally lower than in fixed automation, because the equipment is designed to facilitate product changeover rather than for product specialization. A numerical-control machine tool is a good example of programmable automation. The program is coded in computer memory for each different product style, and the machine tool is controlled by the computer program. Industrial robots are another example.

Flexible automation is an extension of programmable automation. The disadvantage with programmable automation is the time required to reprogram and change over the production equipment for each batch of new product. This is lost production time, which is expensive. In flexible automation, the variety of products is sufficiently limited so that the changeover of the equipment can be done very quickly and automatically. The reprogramming of the equipment in flexible automation is done off-line; that is, the programming is accomplished at a computer terminal without using the production equipment itself. Accordingly, there is no need to group identical products into batches; instead, a mixture of different products can be produced one right after another.

Automated Production Lines

An automated production line consists of a series of workstations connected by a transfer system to move parts between the stations. This is an example of fixed automation, since these lines are typically set up for long production runs, perhaps making millions of product units and running for several years between changeovers. Each station is designed to perform a specific processing operation, so that the part or product is constructed stepwise as it progresses along the line. A raw work part enters at one end of the line, proceeds through each workstation, and emerges at the other end as a completed product. In the normal operation of the line, there is a work part being processed at each station, so that many parts are being processed simultaneously and a finished part is produced with each cycle of the line. The various operations, part transfers, and other activities taking place on an automated transfer line must all be sequenced and coordinated properly for the line to operate efficiently. Modern automated lines are controlled by programmable logic controllers, which are special computers that facilitate connections with industrial equipment (such as automated production lines) and can perform the kinds of timing and sequencing functions required to operate such equipment.

Automated production lines are utilized in many industries, most notably automotive, where they are used for processes such as machining and press working. Machining is a manufacturing process in which metal is removed by a cutting or shaping tool, so that the remaining work part is the desired shape. Machinery and motor components are usually made by this process. In many cases, multiple operations are required to completely shape the part. If the part is mass-produced, an automated transfer line is often the most economical method of production. The many separate operations are divided among the workstations. Transfer lines date back to about 1924.

Pressworking operations involve the cutting and forming of parts from sheet metal. Examples of such parts include automobile body panels, outer shells of major appliances (e.g., laundry machines and ranges), and metal furniture (e.g., desks and file cabinets). More than one processing step is often required to complete a complicated part. Several presses are connected together in sequence by handling mechanisms that transfer the partially completed parts from one press to the next, thus creating an automated pressworking line.

Numerical Control

Numerical control is a form of programmable automation in which a machine is controlled by numbers (and other symbols) that have been coded on punched paper tape or an alternative storage medium. The initial application of numerical control was in the machine tool industry, to control the position of a cutting tool relative to the work part being machined. The NC part program represents the set of machining instructions for the particular part. The coded numbers in the program specify x-y-z coordinates in a Cartesian axis system, defining the various positions of the cutting tool in relation to the work part. By sequencing these positions in the program, the machine tool is directed to accomplish the machining of the part. A position feedback control system is used in most NC machines to verify that the coded instructions have been correctly performed.

Today a small computer is used as the controller in an NC machine tool, and the program is actuated from computer memory rather than punched paper tape. However, initial entry of the program into computer memory is often still accomplished using punched tape. Since this form of numerical control is implemented by computer, it is called computer numerical control, or CNC. Another variation in the implementation of numerical control involves sending part programs over telecommunications lines from a central computer to individual machine tools in the factory, thus eliminating the use of the punched tape altogether. This form of numerical control is called direct numerical control, or DNC.

Computer numerical control: Rows of computer numerical control (CNC) milling machines.

Many applications of numerical control have been developed since its initial use to control machine tools. Other machines using numerical control include component-insertion machines used in electronics assembly, drafting machines that prepare engineering drawings, coordinate measuring machines that perform accurate inspections of parts, and flame cutting machines and similar devices. In these applications, the term numerical control is not always used explicitly, but the operating principle is the same: coded numerical data are employed to control the position of a tool or workhead relative to some object.

To illustrate these alternative applications of numerical control, the component-insertion machine will be considered here. Such a machine is used to position electronic components (e.g., semiconductor chip modules) onto a printed circuit board (PCB). It is basically an x-y positioning table that moves the printed circuit board relative to the part-insertion head, which then places the individual component into position on the board. A typical printed circuit board has dozens of individual components that must be placed on its surface; in many cases, the lead wires of the components must be inserted into small holes in the board, requiring great precision by the insertion machine. The program that controls the machine indicates which components are to be placed on the board and their locations. This information is contained in the product-design database and is typically communicated directly from the computer to the insertion machine.

Automated Assembly

Assembly operations have traditionally been performed manually, either at single assembly workstations or on assembly lines with multiple stations. Owing to the high labour content and high cost of manual labour, greater attention has been given in recent years to the use of automation for assembly work. Assembly operations can be automated using production line principles if the quantities are large, the product is small, and the design is simple (e.g., mechanical pencils, pens, and cigarette lighters). For products that do not satisfy these conditions, manual assembly is generally required.

Automated assembly machines have been developed that operate in a manner similar to machining transfer lines, with the difference being that assembly operations, instead of machining, are performed at the workstations. A typical assembly machine consists of several stations, each equipped with a supply of components and a mechanism for delivering the components into position for assembly. A workhead at each station performs the actual attachment of the component. Typical workheads include automatic screwdrivers, staking or riveting machines, welding heads, and other joining devices. A new component is added to the partially completed product at each workstation, thus building up the product gradually as it proceeds through the line. Assembly machines of this type are considered to be examples of fixed automation, because they are generally configured for a particular product made in high volume. Programmable assembly machines are represented by the component-insertion machines employed in the electronics industry.

Robots in Manufacturing

Today most robots are used in manufacturing operations; the applications can be divided into three categories: (1) material handling, (2) processing operations, and (3) assembly and inspection.

Material-handling applications include material transfer and machine loading and unloading. Material-transfer applications require the robot to move materials or work parts from one location to another. Many of these tasks are relatively simple, requiring robots to pick up parts from one conveyor and place them on another. Other transfer operations are more complex, such as placing parts onto pallets in an arrangement that must be calculated by the robot. Machine loading and unloading operations utilize a robot to load and unload parts at a production machine. This requires the robot to be equipped with a gripper that can grasp parts. Usually the gripper must be designed specifically for the particular part geometry.

In robotic processing operations, the robot manipulates a tool to perform a process on the work part. Examples of such applications include spot welding, continuous arc welding, and spray painting. Spot welding of automobile bodies is one of the most common applications of industrial robots in the United States. The robot positions a spot welder against the automobile panels and frames to complete the assembly of the basic car body. Arc welding is a continuous process in which the robot moves the welding rod along the seam to be welded. Spray painting involves the manipulation of a spray-painting gun over the surface of the object to be coated. Other operations in this category include grinding, polishing, and routing, in which a rotating spindle serves as the robot's tool.

The third application area of industrial robots is assembly and inspection. The use of robots in assembly is expected to increase because of the high cost of manual labour common in these operations. Since robots are programmable, one strategy in assembly work is to produce multiple product styles in batches, reprogramming the robots between batches. An alternative strategy is to produce a mixture of different product styles in the same assembly cell, requiring each robot in the cell to identify the product style as it arrives and then execute the appropriate task for that unit.

The design of the product is an important aspect of robotic assembly. Assembly methods that are satisfactory for humans are not necessarily suitable for robots. Using a screw and nut as a fastening method, for example, is easily performed in manual assembly, but the same operation is extremely difficult for a one-armed robot. Designs in which the components are to be added from the same direction using snap fits and other one-step fastening procedures enable the work to be accomplished much more easily by automated and robotic assembly methods.

Inspection is another area of factory operations in which the utilization of robots is growing. In a typical inspection job, the robot positions a sensor with respect to the work part and determines whether the part is consistent with the quality specifications.

In nearly all industrial robotic applications, the robot provides a substitute for human labour. There are certain characteristics of industrial jobs performed by humans that identify the work as a potential application for robots: (1) the operation is repetitive, involving the same basic work motions every cycle; (2) the operation is hazardous or uncomfortable for the human worker (e.g., spray painting, spot welding, arc welding, and certain machine loading and unloading tasks); (3) the task requires a work part or tool that is heavy and awkward to handle; and (4) the operation allows the robot to be used on two or three shifts.

Flexible Manufacturing Systems

A flexible manufacturing system (FMS) is a form of flexible automation in which several machine tools are linked together by a material-handling system, and all aspects of the system are controlled by a central computer. An FMS is distinguished from an automated production line by its ability to process more than one product style simultaneously. At any moment, each machine in the system may be processing a different part type. An FMS can also cope with changes in product mix and production schedule as demand patterns for the different products made on the system change over time. New product styles can be introduced into production with an FMS, so long as they fall within the range of products that the system is designed to process. This kind of system is therefore ideal when demand for the products is low to medium and there are likely to be changes in demand.

The components of an FMS are (1) processing machines, which are usually CNC machine tools that perform machining operations, although other types of automated workstations such as inspection stations are also possible, (2) a material-handling system, such as a conveyor system, which is capable of delivering work parts to any machine in the FMS, and (3) a central computer system that is responsible for communicating NC part programs to each machine and for coordinating the activities of the machines and the material-handling system. In addition, a fourth component of an FMS is human labour. Although the flexible manufacturing system represents a high level of production automation, people are still needed to manage the system, load and unload parts, change tools, and maintain and repair the equipment.

Computer Process Control

In computer process control, a digital computer is used to direct the operations of a manufacturing process. Although other automated systems are typically controlled by computer, the term computer process control is generally associated with continuous or semicontinuous production operations involving materials such as chemicals, petroleum, foods, and certain basic metals. In these operations the products are typically processed in gas, liquid, or powder form to facilitate flow of the material through the various steps of the production cycle. In addition, these products are usually mass-produced. Because of the ease of handling the product and the large volumes involved, a high level of automation has been accomplished in these industries.

The modern computer process control system generally includes the following: (1) measurement of important process variables such as temperature, flow rate, and pressure, (2) execution of some optimizing strategy, (3) actuation of such devices as valves, switches, and furnaces that enable the process to implement the optimal strategy, and (4) generation of reports to management indicating equipment status, production performance, and product quality. Today computer process control is applied to many industrial operations.

The typical modern process plant is computer-controlled. In one petrochemical plant that produces more than 20 products, the facility is divided into three areas, each with several chemical-processing units. Each area has its own process-control computer to perform scanning, control, and alarm functions. The computers are connected to a central computer in a hierarchical configuration. The central computer calculates how to obtain maximum yield from each process and generates management reports on process performance.

Each process computer monitors up to 2,000 parameters that are required to control the process, such as temperature, flow rate, pressure, liquid level, and chemical concentration. These measurements are taken on a sampling basis; the time between samples varies between 2 and 120 seconds, depending on the relative need for the data. Each computer controls approximately 400 feedback control loops. Under normal operation, each control computer maintains operation of its process at or near optimum performance levels. If process parameters exceed the specified normal or safe ranges, the control computer actuates a signal light and alarm horn and prints a message indicating the nature of the problem for the technician. The central computer receives data from the process computers and performs calculations to optimize the performance of each chemical-processing unit. The results of these calculations are then passed to the individual process computers in the form of changes in the set points for the various control loops.

Substantial economic advantages are obtained from this type of computer control in the process industries. The computer hierarchy is capable of integrating all the data from the many individual control loops far better than humans are able to do, thus permitting a higher level of performance. Advanced control algorithms can be applied by the computer to optimize the process. In addition, the computer is capable of sensing process conditions that indicate unsafe or abnormal operation much more quickly than humans can. All these improvements increase productivity, efficiency, and safety during process operation.

Like the chemical-processing industries, the basic metals industries (iron and steel, aluminum, etc.) have automated many of their processes by computer control. Like the chemical industries, the metals industries deal in large volumes of products, and so there is a substantial economic incentive to invest in automation. However, metals are typically produced in batches rather than continuously, and it is generally more difficult to handle metals in bulk form than chemicals that flow.

An example of computer process control in the metals industry is the rolling of hot metal ingots into final shapes such as coils and strips. This was first done in the steel industry, but similar processing is also accomplished with aluminum and other metals. In a modern steel plant, hot-rolling is performed under computer control. The rolling process involves the forming of a large, hot metal billet by passing it through a rolling mill consisting of one or more sets of large cylindrical rolls that squeeze the metal and reduce its cross section. Several passes are required to reduce the ingot gradually to the desired thickness. Sensors and automatic instruments measure the dimensions and temperature of the ingot after each pass through the rolls, and the control computer calculates and regulates the roll settings for the next pass.

In a large plant, several orders for rolled products with different specifications may be in the mill at any given time. Control programs have been developed to schedule the sequence and rate at which the hot metal ingots are fed through the rolling mills. The production control task of scheduling and keeping track of the different orders requires rapid, massive data gathering and analysis. In modern plants this task has been effectively integrated with the computer control of the rolling mill operations to achieve a highly automated production system.

Computer-integrated Manufacturing

Since about 1970 there has been a growing trend in manufacturing firms toward the use of computers to perform many of the functions related to design and production. The technology associated with this trend is called CAD/CAM, for computer-aided design and computer-aided manufacturing. Today it is widely recognized that the scope of computer applications must extend beyond design and production to include the business functions of the firm. The name given to this more comprehensive use of computers is computer-integrated manufacturing (CIM).

CAD/CAM is based on the capability of a computer system to process, store, and display large amounts of data representing part and product specifications. For mechanical products, the data represent graphic models of the components; for electrical products, they represent circuit information; and so forth. CAD/CAM technology has been applied in many industries, including machined components, electronics products, and equipment design and fabrication for chemical processing. CAD/CAM involves not only the automation of the manufacturing operations but also the automation of elements in the entire design-and-manufacturing procedure.

Computer-aided design (CAD) makes use of computer systems to assist in the creation, modification, analysis, and optimization of a design. The designer, working with the CAD system rather than the traditional drafting board, creates the lines and surfaces that form the object (product, part, structure, etc.) and stores this model in the computer database. By invoking the appropriate CAD software, the designer can perform various analyses on the object, such as heat transfer calculations. The final object design is developed as adjustments are made on the basis of these analyses. Once the design procedure has been completed, the computer-aided design system can generate the detailed drawings required to make the object.

Computer-aided manufacturing (CAM) involves the use of computer systems to assist in the planning, control, and management of production operations. This is accomplished by either direct or indirect connections between the computer and production operations. In the case of the direct connection, the computer is used to monitor or control the processes in the factory. Computer process monitoring involves the collection of data from the factory, the analysis of the data, and the communication of process-performance results to plant management. These measures increase the efficiency of plant operations. Computer process control entails the use of the computer system to execute control actions to operate the plant automatically. Indirect connections between the computer system and the process involve applications in which the computer supports the production operations without actually monitoring or controlling them. These applications include planning and management functions that can be performed by the computer (or by humans working with the computer) more efficiently than by humans alone. Examples of these functions are planning the step-by-step processes for the product, part programming in numerical control, and scheduling the production operations in the factory.

Computer-integrated manufacturing includes all the engineering functions of CAD/CAM and the business functions of the firm as well. These business functions include order entry, cost accounting, employee time records and payroll, and customer billing. In an ideal CIM system, computer technology is applied to all the operational and information-processing functions of the company, from customer orders through design and production (CAD/CAM) to product shipment and customer service. The scope of the computer system includes all activities that are concerned with manufacturing. In many ways, CIM represents the highest level of automation in manufacturing.

Automation and Society

Over the years, the social merits of automation have been argued by labour leaders, business executives, government officials, and college professors. The biggest controversy has focused on how automation affects employment. There are other important aspects of automation, including its effect on productivity, economic competition, education, and quality of life.

Impact on the Individual

Nearly all industrial installations of automation, and in particular robotics, involve a replacement of human labour by an automated system. Therefore, one of the direct effects of automation in factory operations is the dislocation of human labour from the workplace. The long-term effects of automation on employment and unemployment rates are debatable. Most studies in this area have been controversial and inconclusive. Workers have indeed lost jobs through automation, but population increases and consumer demand for the products of automation have compensated for these losses. Labour unions have argued, and many companies have adopted the policy, that workers displaced by automation should be retrained for other positions, perhaps increasing their skill levels in the process. This argument succeeds so long as the company and the economy in general are growing at a rate fast enough to create new positions as the jobs replaced by automation are lost.

Of particular concern for many labour specialists is the impact of industrial robots on the work force, since robot installations involve a direct substitution of machines for humans, sometimes at a ratio of two to three humans per robot. The opposing argument within the United States is that robots can increase productivity in American factories, thereby making these firms more competitive and ensuring that jobs are not lost to overseas companies. The effect of robotics on labour has been relatively minor, because the number of robots in the United States is small compared with the number of human workers. As of the early 1990s, there were fewer than 100,000 robots installed in American factories, compared with a total work force of more than 100 million persons, about 20 million of whom work in factories.

Automation affects not only the number of workers in factories but also the type of work that is done. The automated factory is oriented toward the use of computer systems and sophisticated programmable machines rather than manual labour. Greater emphasis is placed on knowledge-based work and technical skill rather than physical work. The types of jobs found in modern factories include more machine maintenance, improved scheduling and process optimization, systems analysis, and computer programming and operation. Consequently, workers in automated facilities must be technologically proficient to perform these jobs. Professional and semiprofessional positions, as well as traditional labour jobs, are affected by this shift in emphasis toward factory automation.

Impact on Society

Besides affecting individual workers, automation has an impact on society in general. Productivity is a fundamental economic issue that is influenced by automation. The productivity of a process is traditionally defined as the ratio of output units to the units of labour input. A properly justified automation project will increase productivity owing to increases in production rate and reductions in labour content. Over the years, productivity gains have led to reduced prices for products and increased prosperity for society.

A number of issues related to education and training have been raised by the increased use of automation, robotics, computer systems, and related technologies. As automation has increased, there has developed a shortage of technically trained personnel to implement these technologies competently. This shortage has had a direct influence on the rate at which automated systems can be introduced. The shortage of skilled staffing in automation technologies raises the need for vocational and technical training to develop the required work-force skills. Unfortunately the educational system is also in need of technically qualified instructors to teach these subjects, and the laboratory equipment available in schools does not always represent the state-of-the-art technology typically used in industry.

Advantages and Disadvantages of Automation

Advantages commonly attributed to automation include higher production rates and increased productivity, more efficient use of materials, better product quality, improved safety, shorter workweeks for labour, and reduced factory lead times. Higher output and increased productivity have been two of the biggest reasons in justifying the use of automation. Despite the claims of high quality from good workmanship by humans, automated systems typically perform the manufacturing process with less variability than human workers, resulting in greater control and consistency of product quality. Also, increased process control makes more efficient use of materials, resulting in less scrap.

Worker safety is an important reason for automating an industrial operation. Automated systems often remove workers from the workplace, thus safeguarding them against the hazards of the factory environment. In the United States the Occupational Safety and Health Act of 1970 (OSHA) was enacted with the national objective of making work safer and protecting the physical well-being of the worker. OSHA has had the effect of promoting the use of automation and robotics in the factory.

Another benefit of automation is the reduction in the number of hours worked on average per week by factory workers. About 1900 the average workweek was approximately 70 hours. This has gradually been reduced to a standard workweek in the United States of about 40 hours. Mechanization and automation have played a significant role in this reduction. Finally, the time required to process a typical production order through the factory is generally reduced with automation.

Despite the social benefits that might result from retraining displaced workers for other jobs, in almost all cases the worker whose job has been taken over by a machine undergoes a period of emotional stress. In addition to displacement from work, the worker may be displaced geographically. In order to find other work, an individual may have to relocate, which is another source of stress.

Other disadvantages of automated equipment include the high capital expenditure required to invest in automation (an automated system can cost millions of dollars to design, fabricate, and install), a higher level of maintenance needed than with a manually operated machine, and a generally lower degree of flexibility in terms of the possible products as compared with a manual system (even flexible automation is less flexible than humans, the most versatile machines of all).

Also there are potential risks that automation technology will ultimately subjugate rather than serve humankind. The risks include the possibility that workers will become slaves to automated machines, that the privacy of humans will be invaded by vast computer data networks, that human error in the management of technology will somehow endanger civilization, and that society will become dependent on automation for its economic well-being.

These dangers aside, automation technology, if used wisely and effectively, can yield substantial opportunities for the future. There is an opportunity to relieve humans from repetitive, hazardous, and unpleasant labour in all forms. And there is an opportunity for future automation technologies to provide a growing social and economic environment in which humans can enjoy a higher standard of living and a better way of life.

Elements of Automated System

An automated system consists of three basic elements:

- Power to accomplish the process and operate the system.
- A program of instructions to direct the process.
- A control system to actuate the instructions.

The relationship amongst these elements is illustrated in figure. All systems that qualify as being automated include these three basic elements in one form or another.

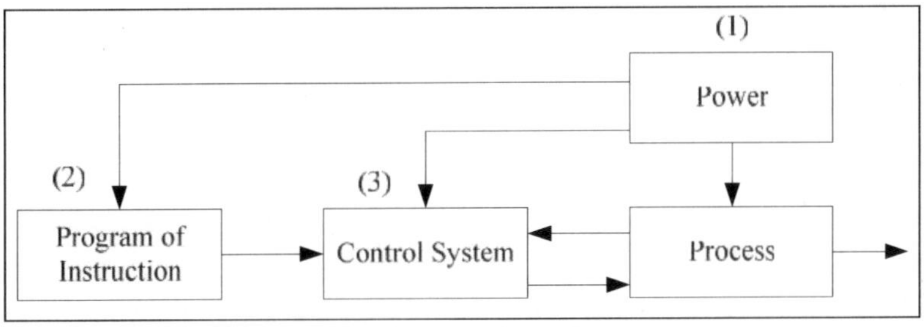

Elements of an automated system.

Power to Accomplish the Automated Process

An automated system is used to operate some process, and power is required to drive the process as well as the controls. The principal source of power in automated systems is electricity. Electric power has many advantages in automated as well as non-automated processes:

- Electrical power is widely available at moderate cost.
- Electrical power can be readily converted to alternative energy forms: mechanical, thermal, light, acoustic, hydraulic, and pneumatic.
- Electrical power at low levels can be used to accomplish functions such as signal, transmission, information processing, and data storage and communication.
- Electrical energy can be stored in long-life batteries for use in locations where an external source of electrical power is not conveniently available.

Power is required in automation for the followings:

- Processing operations.
- Loading and unloading the work unit.
- Material transport between operations.
- Controller unit.
- Power to actuate the control signals.
- Data acquisition and information processing.

Program of Instructions

The actions performed in an automated process are defined by a program of instructions. Each part or product style made in the operation requires one or more processing steps that are unique to that style, These processing steps are performed during a work cycle. A new part is completed during each work cycle (in some manufacturing operations, more than one part is produced during the work cycle; e.g., a plastic injection molding operation may produce multiple parts each cycle using a multiple cavity mold). The particular processing steps for the work cycle are specified in a work cycle program.

Work Cycle Programs

In the simplest automated processes, the work cycle consists of essentially one step, which is to maintain a single process parameter at a defined level. However, the system becomes complicated when the process involves a work cycle consisting of multiple steps with more number of process parameters are required to be controlled. Most discrete part manufacturing operations are in this category.

Process Parameters

Process parameters are inputs to the process such as temperature setting of a furnace, coordinate axis value in a positioning system, valve opened or closed in a fluid flow system, and motor on or off. Process parameters are distinguished from process variables, which are outputs from the process; for example, the actual temperature of the furnace, the actual position of the axis, the actual flow rate of the fluid in the pipe, and the rotational speed of the motor. As our list of examples suggests, the changes in process parameter values may be continuous (gradual changes during the processing step; for example, gradually increasing temperature during a heat treatment cycle) or discrete (stepwise changes; for example, on/off).

The work cycle may include manual steps, where the operator performs certain activities during the work cycle and the automated system performs the rest. A common example is the loading and unloading of parts by the operator to and from a numerical control machine between machining cycles where the machine performs the cutting operation under part program control. Initiation of the cutting operation of each cycle is triggered by the operator activating a "start" button after the part has been loaded.

Decision-making in the Programmed Work Cycle

In automated work cycles the only two features are (1) the number and sequence of processing steps and (2) the process parameter changes in each step. Each work cycle consists of the same steps and associated process parameter changes with no variation from one cycle to the next. The program of instructions is repealed each work cycle without deviation. In fact, many automated manufacturing operations require decisions to be made during the programmed work cycle to cope with variations in the cycle. In many cases, the variations are routine elements of the cycle, and the corresponding instructions for dealing with them are incorporated into the regular part program. These cases include:

- Operator interaction: Although the program of instructions is intended to be carried out without human interaction, the controller unit may require input data from a human operator in order to function.
- Different part or product styles processed by the System: In this instance, the automated system is programmed to perform different work cycles on different part or product styles.
- Variations in the starting work units: In many manufacturing operations the starting work units are not consistent. A good example is a sand casting as the starting work unit in a machining operation. The dimensional variations in the raw castings sometimes necessitate an extra machining pass to bring the machined dimension to the specified value. The part program must be coded to allow for the additional pass when necessary.

In all of these examples, the routine variations can be accommodated in the regular work cycle program. The program can be designed to respond to sensor or operator inputs by executing the appropriate subroutine corresponding to the input. In other cases, the variations in the work cycle are not routine at all. They are infrequent and unexpected, such as the failure of an equipment component. In these instances, the program must include contingency procedures or modifications in the sequence to cope with conditions that lie outside the normal routine.

Control System

The control element of the automated system executes the program of instructions. The control system causes the process to accomplish its defined function which is to carry out some manufacturing operation. The controls in an automated system can be either closed loop or open loop. A closed loop control system, also known as a feedback control system is one in which the output variable is compared with an input parameter, and any difference between the two is used to drive the output into agreement with the input. As shown in figure, a closed loop control system consists of six basic elements: (1) input parameter, (2) process, (3) output variable, (4) feedback sensor. (5) controller and (6) actuator.

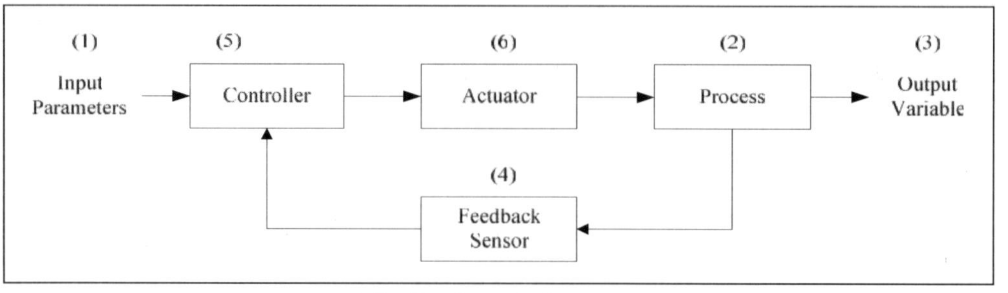

A feedback control system.

The input parameter often referred to as the set point, represents the desired value of the output. The process is the operation or function being controlled. In particular, it is the output variable that is being controlled in the loop. A sensor is used to measure the output variable and close the loop between input and output. Sensors perform the feedback function in a closed loop control system. The controller compares the output with the input and makes the required adjustment in the process to reduce the difference between them. The adjustment is accomplished using one or more actuators, which are the hardware devices that physically carry out the control actions, such as an electric motor or a flow valve. The model in figure shows only one loop, however, most industrial processes require multiple loops, one for each.

In contrast to the closed loop control system, an open loop control system operates without the feedback loop. In this case, the controls operate without measuring the output variable so no comparison is made between the actual value of the output and the desired input parameter. The controller relies on an accurate model of the effect of its actuator on the process variable. With an open loop system, there is always the risk that the actuator will not have the intended effect on the process, and that is the disadvantage of an open loop system. Its advantage is that it is generally simpler and less expensive than a closed loop system. Open loop systems are usually appropriate when the following conditions apply: (1) The actions performed by the control system are simple, (2) the actuating function is very reliable, and (3) any reaction forces opposing the actuation are small enough to have no effect on the actuation. If these characteristics are not applicable, then a closed loop control system may be more appropriate.

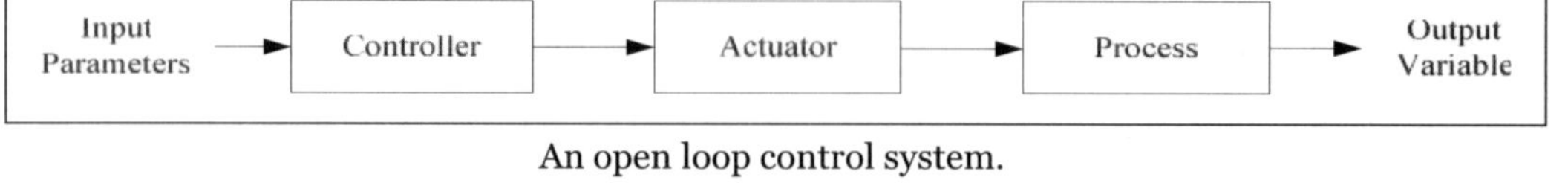

An open loop control system.

Advanced Automation Functions

Advanced automation here refers to sophisticated automated systems, ideally with the additional capability for self-maintenance and repair, mostly requiring little or no human interaction to operate, apart from top-level guidance. Not being reliant on human effort to scale, these systems would hugely magnify our capability for production and decouple human time and effort from industrial productivity, allowing us to create as much of anything that is needed, while releasing people from mindless labour. This situation will arise when automated harvesting of raw materials is combined with automated logistics (already commonplace), automated transport systems, robotic manufacturing and self-maintenance and repair, creating fully automated production of useful goods.

In Western countries many industrial process are becoming highly automated already, but human effort is needed for construction and commissioning as well as maintenance and repair. In developing nations, there is not much automation at all due to labour being so cheap; however this is a great waste of human lives.

These self-repairing systems are based on technologies and knowledge that we already possess. No fictional concepts or unattainable artificial intelligence are required to make this happen. We have the ability today to create systems that provide for the global population's basic needs and far beyond, while minimising our impact on the environment – these two aspects are not mutually exclusive.

Self-maintenance and Repair

If complicated physical systems were able to be serviced and repaired completely automatically there would be many advantages. There would be higher productivity and efficiency without people in the loop - we tend to slow things down and are error prone; also people could be freed up to do something less menial; and the systems could scale quickly when more capacity is needed.

Machines today, such as industrial machinery, are designed to be looked after and serviced by people, and it would likely need artificial intelligence beyond our current capabilities to maintain or repair these systems completely autonomously. However it is feasible to design them from the outset to be maintained autonomously; designed in a modular fashion with components easily removed and replaced by another machine, and embedded wired or wireless sensors giving the ability to diagnose faults on all significant parts.

Many parameters can now be sensed with solid-state sensors, manufactured on tiny silicon chips, which can be embedded within functioning machines. If the signatures from multiple sensors relating to each machine function is known when operating within normal bounds, it provides a method for pin-pointing problems with great accuracy. Vibration, temperature, rotation, pressure, distance, voltage, acceleration and structural integrity as examples. This already happens to a certain extent on machines today such as vehicles but it only applies to a small sub-set of vital components.

What we are talking about here is having multiple micro-sensors within every single component and also scattered throughout structural parts. A little similar to the way humans and animals have pain receptors that let the brain know when a part is damaged. Unlike humans, however, the computer would have a complete blueprint for how it functions, which would allow precise diagnosis, and flawless new parts could be fabricated to replace broken ones.

Operations can be assessed in real-time and if there is a failure then the defective parts, or the relevant sub-assembly, can be replaced. In many instances it may be possible to know a failure is imminent before it actually happens due to an abnormal rise in temperature or vibration for example. With the system containing a full three-dimensional schematic with exact positions and extraction paths for every part, a repair machine can swap the part without requiring any human intervention.

Where would the replacement parts come from? Either held locally in a store, or shipped in via automated transport, or even manufactured on demand via additive fabrication or CNC milling.

So the physical aspects of the machines need to be designed with autonomous replacement in mind, with magnetic, RFID or optical cues that can easily be read by a repair robot, and highly modular design of components allowing them to easily be extracted and replaced. For instance, a gearbox that slots in or out as a single cartridge.

This same principle, by extension could allow these self-maintaining systems to become self-building too, which means that scaling up facilities becomes easy. More autonomous farming equipment required? Just increase the scale of the agricultural manufacturing facilities, and now you have a greater output of agricultural machinery. A little simplistic perhaps, but you get the idea.

The Shape of Sensors to Come

Hitachi's μ-Chip (shown in top image) is 0.4 x 0.4mm in size. With a small antennae (not shown) it is able to transmit a 128 bit ID number. However, in the lower image is a microscope image of Hitachi's new 'RFID powder', currently in development, shown next to a human hair. It is 64 times smaller by area than the original µ-Chip and has the same capability and can be embedded within a sheet of paper. It will not be long before various types of solid-state sensor can be made on a similar scale and embedded within every component of a machine giving unprecedented information relating to its operation.

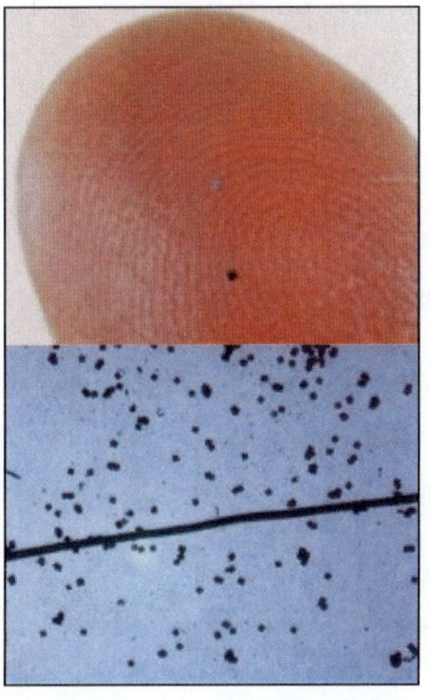

Large Scale Industry

It seems likely that a lot of goods and products made today in factories will increasingly be made using smaller-scale flexible computer-controlled manufacturing methods dispersed across communities and even in homes. Also tasks like water treatment, food production and power generation may well become more distributed and decentralized, again perhaps down to the domestic level.

However larger scale industrial systems are likely to remain for some time to come doing jobs such as mining, material processing and recycling, transport infrastructure and specialised manufacturing and construction.

Increasing Capacity Automatically as Required

A major advantage of having processes almost completely automated is that capacity can easily be scaled up. Just as the manufacturing and construction machinery can be repaired automatically, more manufacturing machinery can also be created by machine, as needed.

This capability means we will be able to do things that are simply not possible at the moment. Mega-scale engineering projects become feasible. If a task is complicated, tedious and a great effort we only need to design the system for the job and let it get on with it. But as these systems become more sophisticated, so will the design tools used to create them. People will be able to interface with these complex systems at ever higher levels of abstraction (although there will be plenty of technically-minded who understand the lower levels too). It will be similar to high-level programming languages hiding the lower levels of code - the individual nuts and bolts will be like the zeros and ones of machine code.

Automated Infrastructure

Automated infrastructure refers to the key infrastructure needed for advanced automation being a fully closed loop. These systems have the capacity to self-maintain and self-repair with little or no human intervention, keeping themselves in the condition and within the parameters set by the engineers. Systems such as transport networks, power generation, water treatment plants, mines, material processing plants, factories and other industrial systems. If these systems are all automated and interface with each other, the means for production for humanity will be ultimately efficient and scalable. These complex systems can be developed using the power of open collaborative design, which has the additional benefit of giving transparency to their development. Having the industrial infrastructure fully automated means it can be easily scaled up to provide everything that the global population requires with ultimate flexibility, and it frees people up to do things that people are good at and want to do.

People in Control

These systems needs to be carefully controlled by people. We will always need to understand how they work, at every level, and they will need to be monitored. Some people wonder whether we will get to a stage where the machines are so sophisticated, and we have relied on them for so long, that no-one will actually know how they work, and therefore we will not be fully in control of them. The reality is that there will always be people interested in this sort of thing - engineers, scientists and geeks in general. They want to know how to make things and understand how they work. There always have been technically-minded people, and have no doubt there always will be.

People will need to improve the designs and make sure they are safe and efficient. We must always remain part of the loop in terms of ultimate control. It is highly likely that we will develop computer-controlled systems more capable at certain tasks than we are, in fact we already have done, but this trend will inevitably continue until there is very little in terms of systems control that can't be done better by a computer. But however sophisticated these systems become they are still just tools for our service — a means to an end. There will always be a threshold where higher-level decisions can only be made with the judgement of people, communities or wider society, and it is important that this threshold should not creep upwards unnoticed over time.

Levels of Automation

The concept of automated systems can be applied to various levels of factory operations. One normally associates automation with the individual production machines. However, the production machine itself is made up of subsystems that may themselves be automated.

For example, a modern numerical control (NC) machine tool is an automated system. However, the NC machine itself is composed of multiple control systems. Similarly, a NC machine is often part of a larger manufacturing system, and the larger system may itself be automated. Thus there are various levels of automation is depicted in figure.

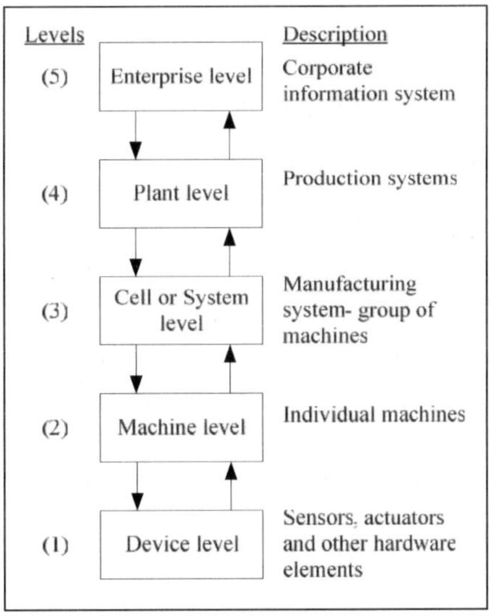

Level of automation.

- Device level: This is the lowest level in automation hierarchy. It includes the actuators, sensors, and other hardware components that comprise the machine level. The devices are combined into the individual control loops of the machine; for example, the feedback control loops for one axis of a CNC machine or one joint of an industrial robot.

- Machine level: Hardware at the device level is assembled into individual machines. Examples include CNC machine tools and similar production equipment, industrial robots, powered conveyors, and automated guided vehicles. Control functions at this level include performing the sequence of steps in the program of instructions in the correct order and making sure that each step is properly executed.
- Cell or system level: This is the manufacturing cell or system level, which operates under instructions from the plant level. A manufacturing cell or system is a group of machines or workstations connected and supported by a material handling system, computer and other equipment appropriate to the manufacturing process. Production lines are included in this level. Likely functions include part dispatching and machine loading, coordination among machines and material handling system, and collecting and evaluating inspection data.
- Plant level: This is the factory or production systems level. It receives instructions from (he corporate information system and translates them into operational plans for production. Likely functions include: order processing, process planning, inventory control, purchasing, material requirements planning, shop floor control, and quality control.
- Enterprise level: This is the highest level consisting of the corporate information system. It is concerned with all of the functions necessary to manage the company: marketing and sales, accounting, design, research, aggregate planning, and master production scheduling.

Automated Testing

Manual Testing is performed by a human sitting in front of a computer carefully executing the test steps.

Automation Testing means using an automation tool to execute your test case suite.

The automation software can also enter test data into the System Under Test, compare expected and actual results and generate detailed test reports. Test Automation demands considerable investments of money and resources.

Successive development cycles will require execution of same test suite repeatedly. Using a test automation tool, it's possible to record this test suite and re-play it as required. Once the test suite is automated, no human intervention is required. This improved ROI of Test Automation. The goal of Automation is to reduce the number of test cases to be run manually and not to eliminate Manual Testing altogether.

Automated software testing is important due to the following reasons:

- Manual Testing of all workflows, all fields, all negative scenarios is time and money consuming.
- It is difficult to test for multilingual sites manually.
- Automation does not require Human intervention. You can run automated test unattended (overnight).
- Automation increases the speed of test execution.
- Automation helps increase Test Coverage.
- Manual Testing can become boring and hence error-prone.

Test cases to be automated can be selected using the following criterion to increase the automation ROI:

- High Risk - Business Critical test cases.
- Test cases that are repeatedly executed.
- Test Cases those are very tedious or difficult to perform manually.
- Test Cases which are time-consuming.

The following category of test cases are not suitable for automation:

- Test Cases that are newly designed and not executed manually at least once.
- Test Cases for which the requirements are frequently changing.
- Test cases which are executed on an ad-hoc basis.

Automated Testing Process

Following steps are followed in an Automation Process:

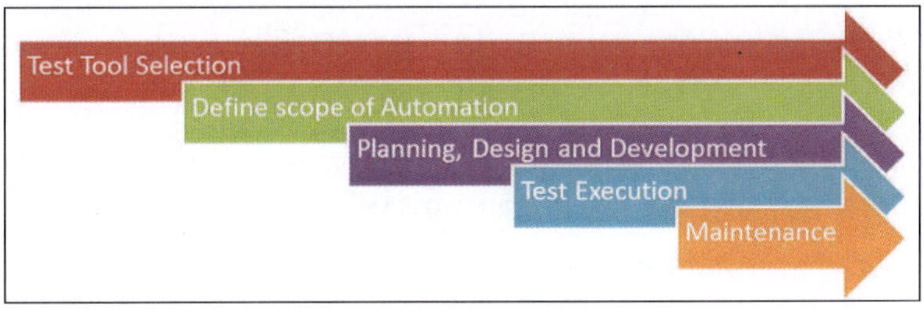

Test Tool Selection

Test Tool selection largely depends on the technology the Application Under Test is built on. For instance, QTP does not support Informatica. So QTP cannot be used for testing Informatica applications. It's a good idea to conduct a Proof of Concept of Tool on AUT.

Scope of Automation

The scope of automation is the area of your Application Under Test which will be automated. Following points help determine scope:

- The features those are important for the business.
- Scenarios which have a large amount of data.
- Common functionalities across applications.
- Technical feasibility.
- The extent to which business components are reused.
- The complexity of test cases.
- Ability to use the same test cases for cross-browser testing.

Planning, Design and Development

During this phase, you create an Automation strategy & plan, which contains the following details:

- Automation tools selected.
- Framework design and its features.
- In-Scope and Out-of-scope items of automation.
- Automation testbed preparation.
- Schedule and Timeline of scripting and execution.
- Deliverables of Automation Testing.

Test Execution

Automation Scripts are executed during this phase. The scripts need input test data before there are set to run. Once executed they provide detailed test reports.

Execution can be performed using the automation tool directly or through the Test Management tool which will invoke the automation tool.

Example: Quality center is the Test Management tool which in turn it will invoke QTP for execution of automation scripts. Scripts can be executed in a single machine or a group of machines. The execution can be done during the night, to save time.

Maintenance

As new functionalities are added to the System Under Test with successive cycles,

Automation Scripts need to be added, reviewed and maintained for each release cycle. Maintenance becomes necessary to improve the effectiveness of Automation Scripts.

Framework for Automation

A framework is set of automation guidelines which help in:

- Maintaining consistency of Testing.
- Improves test structuring.
- Minimum usage of code.
- Less Maintenance of code.
- Improve re-usability.
- Non-Technical testers can be involved in code.
- The training period of using the tool can be reduced.
- Involves Data wherever appropriate.

There are four types of frameworks used in automation software testing:

- Data Driven Automation Framework.
- Keyword Driven Automation Framework.
- Modular Automation Framework.
- Hybrid Automation Framework.

Automation Tool Best Practices

To get maximum ROI of automation, observe the following:

- The scope of Automation needs to be determined in detail before the start of the project. This sets expectations from Automation right.
- Select the right automation tool: A tool must not be selected based on its popularity, but it's fit to the automation requirements.
- Choose an appropriate framework.
- Scripting Standards- Standards have to be followed while writing the scripts for Automation. Some of them are:
    - Create uniform scripts, comments, and indentation of the code.
    - Adequate Exception handling - How error is handled on system failure or unexpected behavior of the application.

    - User-defined messages should be coded or standardized for Error Logging for testers to understand.

- Measure metrics: Success of automation cannot be determined by comparing the manual effort with the automation effort but by also capturing the following metrics:
    - Percent of defects found.
    - The time required for automation testing for each and every release cycle.
    - Minimal Time is taken for release.
    - Customer Satisfaction Index.
    - Productivity improvement.

The above guidelines if observed can greatly help in making your automation successful.

Benefits of Automation Testing

Following are benefits of automated testing:

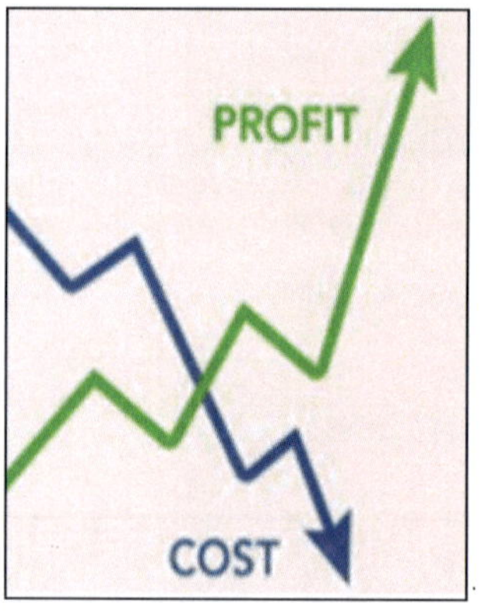

- 70% faster than the manual testing.
- Wider test coverage of application features.
- Reliable in results.
- Ensure Consistency.
- Saves Time and Cost.
- Improves accuracy.
- Human Intervention is not required while execution.
- Increases Efficiency.

- Better speed in executing tests.
- Re-usable test scripts.
- Test frequently and thoroughly.
- More cycle of execution can be achieved through automation.
- Early time to market.

Different types of software testing that can be automated:

- Smoke Testing,
- Unit Testing,
- Integration Testing,
- Functional Testing,
- Keyword Testing,
- Regression Testing,
- Data Driven Testing,
- Black Box Testing.

Choosing an Automation Tool

Selecting the right tool can be a tricky task. Following criterion will help you select the best tool for your requirement:

- Environment Support,
- Ease of use,
- Testing of Database,
- Object identification,
- Image Testing,
- Error Recovery Testing,
- Object Mapping,
- Scripting Language Used,
- Support for various types of test - including functional, test management, mobile, etc.
- Support for multiple testing frameworks,

- Easy to debug the automation software scripts,
- Ability to recognize objects in any environment,
- Extensive test reports and results,
- Minimize training cost of selected tools.

Tool selection is one of biggest challenges to be tackled before going for automation. First, Identify the requirements, explore various tools and its capabilities, set the expectation from the tool and go for a Proof Of Concept.

Automation Testing Tools

There are tons of Functional and Regression Testing Tools available in the market. Here are best tools certified by our experts:

Ranorex Studio

Ranorex Studio is an all-in-one tool for automating functional UI tests, regression tests, data-driven tests and much more. Ranorex Studio includes an easy to use click-and-go interface to automate tests for web, desktop, and mobile applications.

Features:

- Functional UI and end-to-end testing on desktop, web, and mobile.
- Cross-browser testing.
- SAP, ERP, Delphi and legacy applications.
- iOS and Android.
- Run tests locally or remotely, in parallel or distribute on a Selenium Grid.
- Robust reporting.

Mabl

Mabl delivers scriptless end-to-end test automation, integrated with your delivery pipeline, so you can focus on improving your app.

Features:

- Proprietary machine learning models automatically identify and surface application issues.
- Tests are automatically repaired when UI changes.
- Automated regression insights on every build.

Selenium

It is a software testing tool used for Regression Testing. It is an open source testing tool that provides playback and recording facility for Regression Testing. The Selenium IDE only supports Mozilla Firefox web browser.

- It provides the provision to export recorded script in other languages like Java, Ruby, RSpec, Python, C#, etc.
- It can be used with frameworks like JUnit and TestNG.
- It can execute multiple tests at a time.
- Autocomplete for Selenium commands that are common.
- Walkthrough tests.
- Identifies the element using id, name, X-path, etc.
- Store tests as Ruby Script, HTML, and any other format.
- It provides an option to assert the title for every page.
- It supports selenium user-extensions.js file.
- It allows inserting comments in the middle of the script for better understanding and debugging.

QTP (MicroFocus UFT)

QTP is widely used for functional and regression testing, it addresses every major software application and environment. To simplify test creation and maintenance, it uses the concept of keyword driven testing. It allows the tester to build test cases directly from the application.

- It is easier to use for a non-technical person to adapt to and create working test cases.
- It fix defects faster by thoroughly documenting and replicating defects for developer.
- Collapse test creation and test documentation at a single site.
- Parameterization is easy than WinRunner.
- QTP supports .NET development environment.
- It has better object identification mechanism.
- It can enhance existing QTP scripts without "Application Under Test" is available, by using the active screen.

Rational Functional Tester

It is an Object-Oriented automated Functional Testing tool that is capable of performing automated functional, regression, data-driven testing and GUI testing. The main features of this tool are:

- It supports a wide range of protocols and applications like Java, HTML, NET, Windows, SAP, Visual Basic, etc.
- It can record and replay the actions on demand.
- It integrates well with source control management tools such as Rational Clear Case and Rational Team Concert integration.
- It allows developers to create keyword associated script so that it can be re-used.
- Eclipse Java Developer Toolkit editor facilitates the team to code test scripts in Java with Eclipse.
- It supports custom controls through proxy SDK (Java/.Net).
- It supports version control to enable parallel development of test scripts and concurrent usage by geographically distributed team.

Watir

It is an open source testing software for regression testing. It enables you to write tests that are easy to read and maintain. Watir supports only internet explorer on windows while Watir webdriver supports Chrome, Firefox, IE, Opera, etc.

- It supports multiple browsers on different platforms.
- Rather than using proprietary vendor script, it uses a fully-featured modern scripting language Ruby.
- It supports your web app regardless of what it is developed in.

Silk Test

Silk Test is designed for doing functional and regression testing. For e-business application, silk test is the leading functional testing product. It is a product of Segue Software takeover by Borland in 2006. It is an object-oriented language just like C++. It uses the concept of an object, classes, and inheritance. Its main feature includes:

- It consists of all the source script files.
- It converts the script commands into GUI commands. On the same machine, commands can be run on a remote or host machine.

- To identify the movement of the mouse along with keystrokes, Silktest can be executed. It can avail both playback and record method or descriptive programming methods to get the dialogs.
- It identifies all controls and windows of the application under test as objects and determines all of the attributes and properties of each window.

Benefits of Automated Testing

Automated testing can boost a QA team's efficiency. Some benefits include:

- Higher accuracy,
- Better reporting capabilities,
- Increased coverage,
- Enhanced resource efficiency,
- Improved bug detection,
- Increased reusability.

When a software tester manually checks a system, he can make a mistake, especially when an application contains hundreds to thousands of lines of code. Automation helps the QA team avoid these human errors in application testing and executes checks in a faster time frame than if it were done in person.

Some test automation tools have reporting capabilities that log each test script in order to show users the status of every test. A tester can then compare the results with other reports to assess how the software operates compared to expectations and requirements.

Overall, automated testing enables staff to avoid manual tests and focus on other project priorities. A QA team can reuse automated test scripts to ensure each check executes the same way every time. Additionally, automated testing helps a team quickly find bugs in the early stages of development, which can reduce overall working hours and project costs.

Misconceptions about Automated Testing

Misconceptions about automated testing include:

- Automated testing provides developers with more free time. In reality, automated testing gives developers more time to focus on larger issues in the development process.
- Automated testing is superior to manual testing. Automated and manual testing both have their advantages and the most comprehensive understanding of an application will come from utilizing both techniques.

- Automated testing discourages human interaction. In reality, automated testing can enhance conversation by providing new channels to communicate through.
- Automated testing is too expensive. It is the true the initial investment may be costly, but, over time, the benefits of the method help it pay for itself by reducing the cost of code revisions and manually repeating tests.

How does Automated Testing Work?

An organization implements test automation with a framework that includes common practices, testing tools and standards. Data-driven and keyword-driven test automation frameworks are common, as are frameworks for linear scripting and modular testing.

The linear scripting framework suits small applications because it enables the use of a test script with little planning, but does not support reusable scripts. In modular testing frameworks, a software tester creates scripts as small, independent tests to reduce redundancy, but this process typically takes more time to set up.

Data-driven frameworks enable software testers to create scripts that work for multiple data sets and provide wide quality coverage with fewer tests than modular options. Keyword-driven testing frameworks use table formats to define keywords for each function and execution method; software testers without extensive programming knowledge can work with the keywords to create test scripts. Hybrid-driven frameworks combine two or more practices to have the benefits of both.

Open source test automation tools and frameworks include Selenium, Robotium and Cypress. Selenium can automate and run test parameters across multiple web browsers and in various programming languages -- such as C#, Java and Python. Robotium helps testers write automatic user acceptance, function and system tests for Android devices. Cypress covers end-to-end, integration and unit tests, all within a browser. Cypress allows access to distributed object models in the browser and provides a debugger for further tests.

Automated Testing Best Practices

Automated testing is most beneficial when applied to:

- Tests that are performed on different hardware or software configurations or platforms.
- Repetitive tests that are used for various builds.
- Tests with multiple data sets.
- Tests that is impossible to perform manually.
- Tests that are too laborious and time consuming when performed manually.

- Tests on frequently utilized functionality that introduce conditions which elevate risk.
- Tests that frequently generate human error.

Other best practices include:

- Testing the software early and frequently.
- Choosing the correct automated testing tool.
- Creating automated tests that can resists changes in the user interface (UI).
- Separating the automated testing efforts.

Continuous Testing

Organizations typically include automated tests in a continuous testing strategy, which conducts code checks at every step in the software development and delivery pipeline. Continuous and automated tests help organizations reduce performance bottlenecks because the pace of work is ongoing rather than start-and-stop. For example, an organization might release software changes every few hours with automated and continuous testing, rather than every few days with a more manual and gated system.

Continuous delivery and continuous integration pipelines utilize automated tests and deployment processes that enable developers to decide to deploy code when it is ready, as opposed to when the system is available to deploy it. CI involves frequent and isolated code changes, as well as immediate testing at each stage of completion before the CI pipeline adds an update to a larger codebase. CD enables executable code updates to go live in staging or production environments; typically any commit that passes automated integration or other forms of big-picture tests is a valid candidate for release.

Automated Testing and Unit Testing

A unit test is a different software testing method that can be combined with automated testing. Unit testing examines the smallest part of an application in order to ensure functionality. Sometimes, this includes scanning every line of code as a separate piece instead of a part of the whole application. While this can help prevent bugs, it limits the assessment of the overall solution.

When unit testing is performed manually, it can be extremely time consuming and can increase the risk of human errors. Furthermore, manual unit testing removes the collaborative and extensive approach to software development that has be popularized by DevOps culture.

By automating unit tests, multiple test cases can be run as each line of code is written. This ability provides developers with an enhanced understanding of the software's overall integrity and the potential value to end users as it is being developed. Furthermore, the previously discussed benefits of automated testing can be applied to the automation of unit tests: the risk of human error reduces drastically and the time it takes to repeatedly run the tests significantly decreases.

Automated Testing vs. Manual Testing

Manual testing is the exact opposite of automated testing; it involves humans writing and performing all tests on the software. While this extra labor may seem like a disadvantage, it enables developers to benefit from the ability to draw insights from the examination of each step of the process since they are required to go through the software via SQL and log analysis, testing usage and input combinations, comparing collected results to the projected behavior and recording all results.

In contrast, once the test is written, automated testing removes the focus on all the middles steps and instead focuses on delivering the end result. However, this allows tests to be repeatedly performed without the help of developers, thus facilitating continuous testing. In contrast, manual testing requires developers to constantly replicate each step of the process for any test that needs to be repeated on a specified area.

Furthermore, automated testing is frequently used after the software has been developed to run longer tests that were avoided during the initial manual testing. When automated, these lengthy tests can run unattended on multiple computers with various configurations.

References

- Automation, technology: britannica.com, Retrieved 16 July, 2019
- Advanced-automation: adciv.org, Retrieved 18 June, 2019
- Automation-testing: guru99.com, Retrieved 05 May, 2019
- Automated-software-testing: searchsoftwarequality.techtarget.com, Retrieved 08 August, 2019

2

Understanding Computer-Integrated Manufacturing

Computer-Integrated manufacturing deals with the application of computer-controlled machinery and software for automation in manufacturing processes. A few of its types are computer-aided design, computer-aided manufacturing, computer-aided process planning, etc. This chapter discusses the varied types of computer-integrated manufac-turing in detail.

Computer-Integrated Manufacturing (CIM) encompasses the entire range of product development and manufacturing activities with all the functions being carried out with the help of dedicated software packages. The data required for various functions are passed from one application software to another in a seamless manner. For example, the product data is created during design. This data has to be transferred from the modeling software to manufacturing software without any loss of data. CIM uses a common database wherever feasible and communication technologies to integrate design, manufacturing and associated business functions that combine the automated segments of a factory or a manufacturing facility. CIM reduces the human component of manufacturing and thereby relieves the process of its slow, expensive and error-prone component. CIM stands for a holistic and methodological approach to the activities of the manufacturing enterprise in order to achieve vast improvement in its performance.

This methodological approach is applied to all activities from the design of the product to customer support in an integrated way, using various methods, means and techniques in order to achieve production improvement, cost reduction, fulfillment of scheduled delivery dates, quality improvement and total flexibility in the manufacturing system. CIM requires all those associated with a company to involve totally in the process of product development and manufacture. In such a holistic approach, economic, social and human aspects have the same importance as technical aspects. CIM also encompasses the whole lot of enabling technologies including total quality management, business process reengineering, concurrent engineering, workflow automation, enterprise resource planning and flexible manufacturing.

Manufacturing industries strive to reduce the cost of the product continuously to remain competitive in the face of global competition. In addition, there is the need to improve the quality and performance levels on a continuing basis. Another important requirement is on time delivery. In the context of global outsourcing and long supply chains cutting across several international borders, the task of continuously reducing delivery times is really an arduous task. CIM has several software tools to address the above needs.

The challenge before the manufacturing engineers is illustrated in figure.

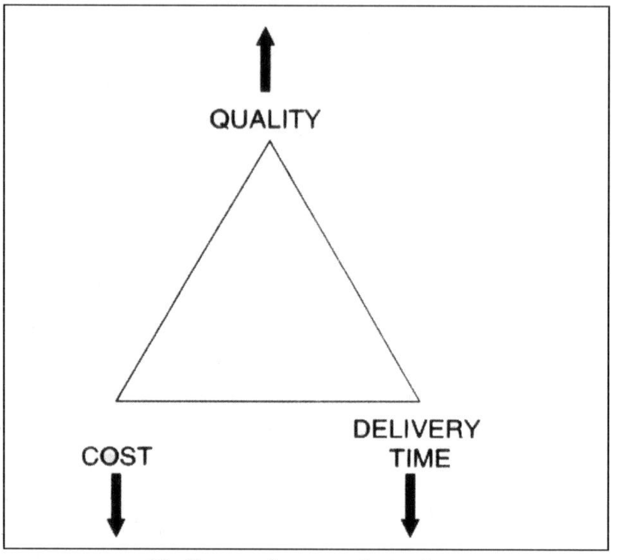

Challenges in manufacturing.

Manufacturing engineers are required to achieve the following objectives to be competitive in a global context.

- Reduction in inventory,
- Lower the cost of the product,
- Reduce waste,
- Improve quality,
- Increase flexibility in manufacturing to achieve immediate and rapid response to:
    - Product changes,
    - Production changes,
    - Process change,
    - Equipment change,
    - Change of personnel.

CIM technology is an enabling technology to meet the above challenges to the manufacturing.

Cim Hardware and CIM Software

CIM Hardware comprises the following:

- Manufacturing equipment such as CNC machines or computerized work centers, robotic work cells, DNC/FMS systems, work handling and tool handling devices, storage devices, sensors, shop floor data collection devices, inspection machines etc.
- Computers, controllers, CAD/CAM systems, workstations / terminals, data entry terminals, bar code readers, RFID tags, printers, plotters and other peripheral devices, modems, cables, connectors etc.

CIM software comprises computer programmes to carry out the following functions:

- Management Information System,
- Sales,
- Marketing,
- Finance,
- Database Management,
- Modeling and Design,
- Analysis,
- Simulation,
- Communications,
- Monitoring,
- Production Control,
- Manufacturing Area Control,
- Job Tracking,
- Inventory Control,
- Shop Floor Data Collection,
- Order Entry,
- Materials Handling,
- Device Drivers,

- Process Planning,
- Manufacturing Facilities Planning,
- Work Flow Automation,
- Business Process Engineering,
- Network Management,
- Quality Management.

Nature and Role of the Elements of CIM System

Nine major elements of a CIM system are in figure they are:

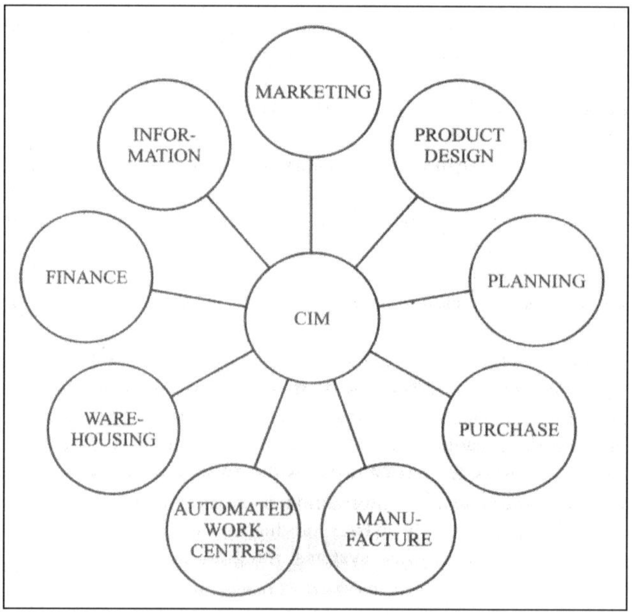

Major elements of CIM systems.

- Marketing: The need for a product is identified by the marketing division. The specifications of the product, the projection of manufacturing quantities and the strategy for marketing the product are also decided by the marketing department. Marketing also works out the manufacturing costs to assess the economic viability of the product.
- Product Design: The design department of the company establishes the initial database for production of a proposed product. In a CIM system this is accomplished through activities such as geometric modeling and computer aided design while considering the product requirements and concepts generated by the creativity of the design engineer. Configuration management is an important activity in many designs. Complex designs are usually carried out by several teams

working simultaneously, located of ten in different parts of the world. The design process is constrained by the costs that will be incurred in actual production and by the capabilities of the available production equipment and processes. The design process creates the database required to manufacture the part.

- Planning: The planning department takes the database established by the design department and enriches it with production data and information to produce a plan for the production of the product. Planning involves several subsystems dealing with materials, facility, process, tools, manpower, capacity, scheduling, outsourcing, assembly, inspection, logistics etc. In a CIM system, this planning process should be constrained by the production costs and by the production equipment and process capability, in order to generate an optimized plan.
- Purchase: The purchase departments is responsible for placing the purchase orders and follow up, ensure quality in the production process of the vendor, receive the items, arrange for inspection and supply the items to the stores or arrange timely delivery depending on the production schedule for eventual supply to manufacture and assembly.
- Manufacturing Engineering: Manufacturing Engineering is the activity of carrying out the production of the product, involving further enrichment of the database with performance data and information about the production equipment and processes. In CIM, this requires activities like CNC programming, simulation and computer aided scheduling of the production activity. This should include online dynamic scheduling and control based on the real time performance of the equipment and processes to assure continuous production activity. Often, the need to meet fluctuating market demand requires the manufacturing system flexible and agile.
- Factory Automation Hardware: Factory automation equipment further enriches the database with equipment and process data, resident either in the operator or the equipment to carry out the production process. In CIM system this consists of computer controlled process machinery such as CNC machine tools, flexible manufacturing systems (FMS), Computer controlled robots, material handling systems, computer controlled assembly systems, flexibly automated inspection systems and so on.
- Warehousing: Warehousing is the function involving storage and retrieval of raw materials, components, finished goods as well as shipment of items. In today's complex outsourcing scenario and the need for just-in-time supply of components and subsystems, logistics and supply chain management assume great importance.
- Finance: Finance deals with the resources pertaining to money. Planning of investment, working capital, and cash flow control, realization of receipts,

accounting and allocation of funds are the major tasks of the finance departments.

- Information Management: Information Management is perhaps one of the crucial tasks in CIM. This involves master production scheduling, database management, communication, manufacturing systems integration and management information systems.

The computer and automated systems association of the society of Manufacturing Engineers (CASA/SEM) defines CIM is the integration of total manufacturing enterprise by using integrated systems and data communication coupled with new managerial philosophies that improve organizational and personnel efficiency.

CIM is recognized as Islands of Automation. They are;

- CAD/CAM/CAE/GT.
- Manufacturing Planning and Control.
- Factory Automation.
- General Business Management.

CASA/SME's CIM Wheel is as shown in figure.

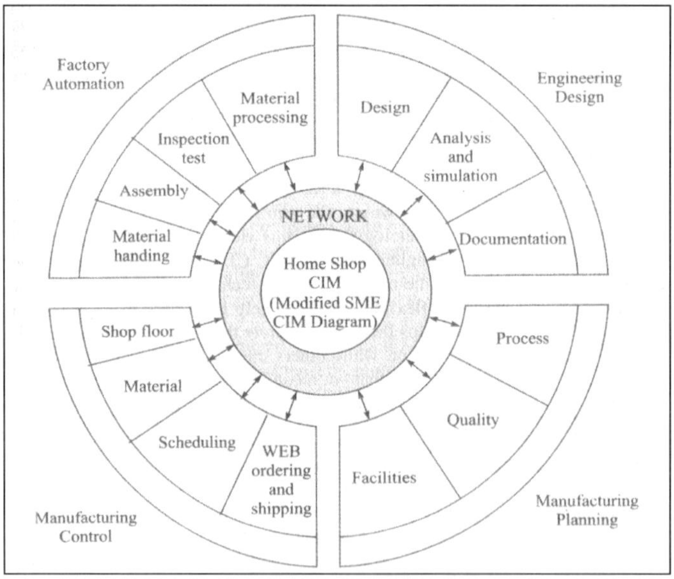

CASA/SME's CIM Wheel.

Conceptual Model of Manufacturing

The computer has had and continues to have a dramatic impact on the development of production automation technologies. Nearly all modern production systems are implemented today using computer systems. The term computer-integrated manufacturing

(CIM) has been coined to denote the pervasive use of computers to design the products, plan the production, control the operations, and perform the various business related functions needed in a manufacturing firm. CAD/CAM (computer-aided design and computer-aided manufacturing) is another term that is used almost synonymously with CIM.

Let us attempt to define the relationship between automation and CIM by developing a conceptual model of manufacturing. In a manufacturing firm, the physical activities related to production that take place in the factory can be distinguished from the information-processing activities, such as product design and production planning, that usually occur in an office environment. The physical activities include all of the manufacturing processing, assembly, material handling, and inspections that are performed on the product. These operations come in direct contact with the product during manufacture. They touch the product. The relationship between the physical activities and the information processing activities in our model is depicted in figure. Raw materials flow in one end of the factory and finished products flow out the other end. The physical activities (processing, handling, etc.) take place inside the factory. The information-processing functions form a ring that surrounds the factory, providing the data and knowledge required to produce the product successfully. These information-processing functions include (1) certain business activities (e.g., marketing and sales, order entry, customer billing, etc.), (2) product design, (3) manufacturing planning, and (4) manufacturing control. These four functions form a cycle of events that must accompany the physical production activities but which do not directly touch the product.

Now consider the difference between automation and CIM. Automation is concerned with the physical activities in manufacturing. Automated production systems are designed to accomplish the processing, assembly, material handling, and inspecting activities with little or no human participation. By comparison, computer-integrated manufacturing is figure.

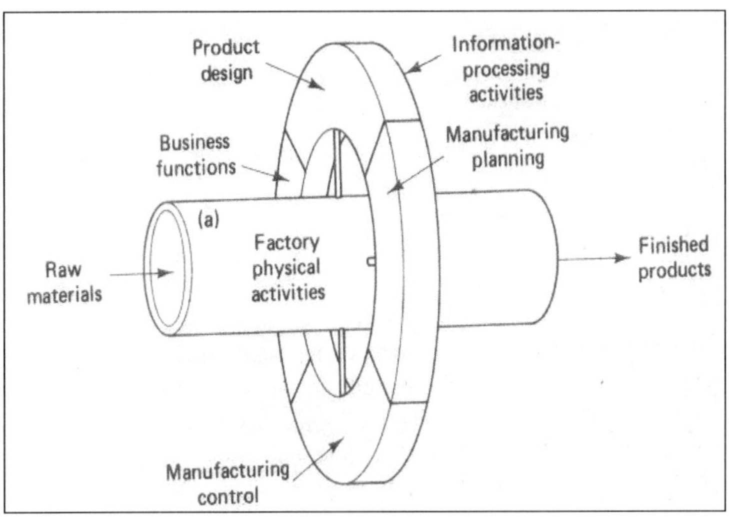

In the figure Model of manufacturing, showing (a) the factory as a processing pipeline where the physical manufacturing activities are performed, and (b) the information-processing activities that support manufacturing as a ring that surrounds the factory concerned more with the information-processing functions that are required to support the production operations. CIM involves the use of computer systems to perform the four types of information-processing functions. Just as automation deals with the physical activities, CIM deals with automating the information-processing activities in manufacturing.

CIM Architecture

A CIM architecture is an information systems structure that enables industrial enterprises integrate information and business processes It accomplishes this first by establishing the direction integration will take; and second, by defining the interfaces between the users and the providers of this integration function.

The chart illustrates how CIM architecture answers the enterprise's integration needs. CIM architecture provides a core of common services. These services support every other area of the enterprise—from its common support functions to its highly specialized business processes.

Three Key Building Blocks

The information environment of an industrial enterprise is subject to frequent changes in systems configuration and technologies. CIM architecture can offer a flexible structure that enables it to react to these changes. This structure relies on a number of modular elements that allow systems to change more easily to grow along with enterprise needs.

The modular elements that give CIM architecture its flexible structure are based on three key building blocks:

- Communications: The communication and distribution of data.
- Data management: The definition, storage and use of data
- Presentation: The presentation of this data to people and devices throughout the enterprise.

Utilizing these building blocks, CIM architecture can provide a consistent base for integrating the enterprise's product, processes and business data. It can define the structure of the hardware, software and services required to support the enterprise's complex requirements. And it can translate this information into a form that can be used by the enterprise's people, devices and applications.

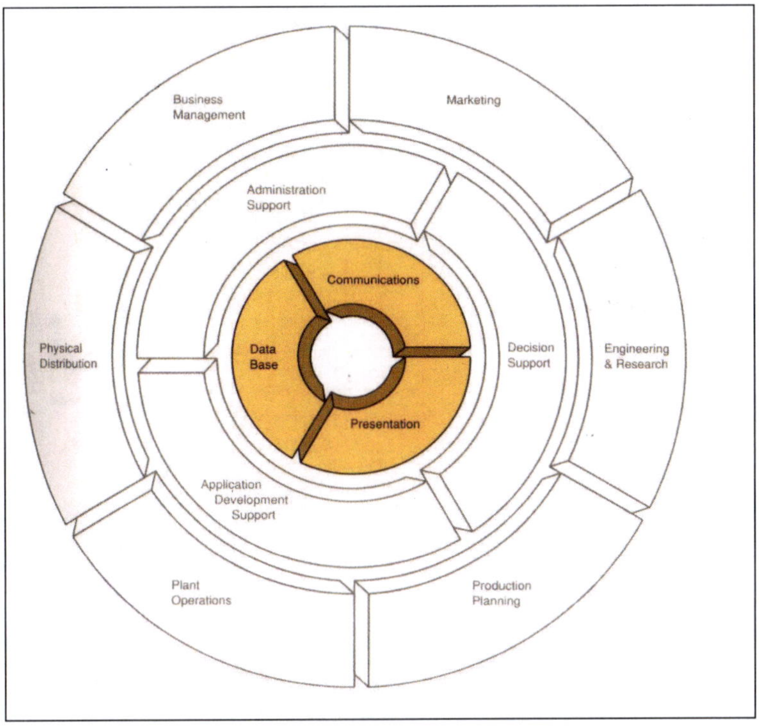

CIM Architecture Elements.

Communications in the CIM Environment

Communications or the delivery of enterprise data to people, systems and devices is a critical aspect of CIM architecture. This is because today's industrial environment brings together a wide range of computer systems, technologies, system architectures, operating systems and applications. This range makes it increasingly difficult for people and machines to communicate with each other especially when they describe and format data differently.

IBM has long recognized this need to communicate data across multiple environments. Our first response was developing Systems Network Architecture (SNA) in the 1970's. SNA supports communication among different IBM systems, and over the years ft has become the standard for host communications in many industrial companies.

However, in the CIM environment communications must be even more integrated. It must expand beyond individual areas, throughout the entire enterprise, and beyond—to customers to vendors and to sub-contractors.

Communications in the CIM environment will involve a wide range of data transfer, from large batches of engineering or planning data to single-bit messages from a plant floor device. Many connectivity types and proto cols must be supported to enable the enterprise's people, systems and devices to communicate. This is especially true in cases where response time is critical, such as during process alerts.

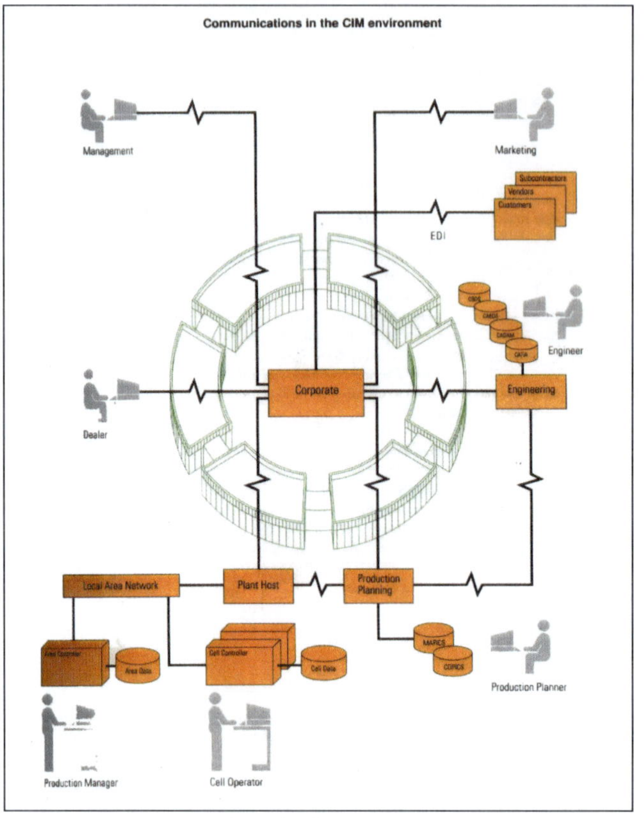

Communications in the CIM environment.

Plant Floor Communications

Let's examine one area where communications can be extremely challenging the plant floor. This is due to the wide range of manufacturing and computer equipment that has been used to manage the various production tasks over the decades.

IBM solution for communicating across these systems is the IBM Plant Floor Series, a set of software products. One of these products, Distributed Automation Edition (DAE), is a systems enabler designed to provide communications functions that can be utilized by plant floor applications. These functions include:

- Defining and managing networks,
- Making logical device assignments,
- Managing a program library,
- Queuing and routing messages,
- Establishing alert procedures,
- Monitoring work-cell status.

These functions enable Distributed Automation Edition to assist manufacturing engineers as they select or develop application programs to control work-cell operations and provide communications capabilities between area- and plant-level systems.

DAE supports several communications protocols to meet the needs of a variety of enterprises. For example, it supports SNA for connections to plant host systems and the IBM PC-Network as well as the IBM Token- Ring protocol and Manufacturing Automated Protocol (MAP) for plant floor communications. MAP is the evolving plant floor communications industry standard, adopted by the International Standards Organization (ISO) for communications among systems provided by different vendors.

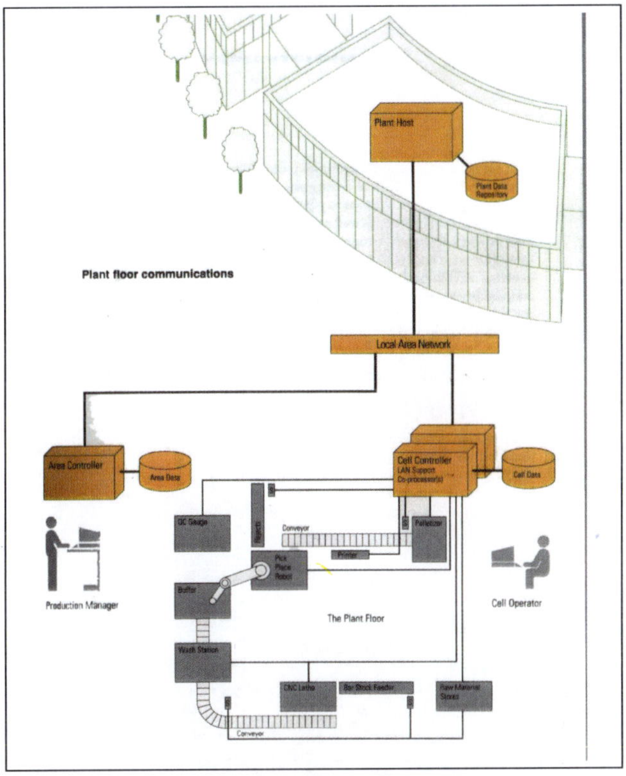

Plant Floor Communications.

Managing Data in the CIM Environment

The second building block of CIM architecture is data management. This includes how data is defined, how different data elements are related, where data is stored, and who has access to that data. Data management is particularly critical in today's industrial environment, since there are so many different data bases, formats, and storage and access techniques.

Standards are evolving. For example, Structured Query Language (SQL) provides a consistent way for relational data base applications and users to acce5s a data base. Unfortunately, there is a significant amount of data that exists today in other data base technologies that is not accessible by current standards that where data management within a CIM architecture can help.

Data management defines and records the location of the data created and used by the enterprise's business functions. Data management also means enabling users to get the data they need—without having to know where this data is located.

Relationships among several data elements must be known if data is to be shared by users and applications. In addition, other data attributes are important when sharing data. These include the type of data (text, graphics, image), its business status (working. review, completed), and the source of this data (person, application, or machine).

In CIM architecture, data management can be accomplished through three individual storage functions:

- The data repository,
- The enterprise data store,
- The local data files.

Some of the key data management functions—the repository, for example—are already being implemented by the Consolidated Design File (CDF) established through the IBM Data Communication Service (DCS).

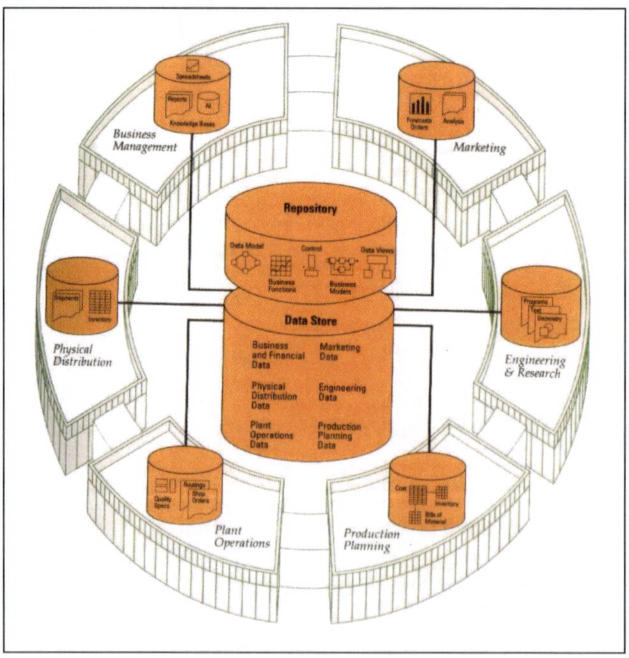

Managing Data in CIM Environment.

The Consolidated Design File operates on a relational data base and is built on SQL. One example of its use is as an engineering data base to integrate CAD/CAM applications with the business needs of the engineering management function. In this environment, IBM's DCS/ CDF provide the following repository functions:

- Transforming data to a user-selected format,
- Storing CAD/CAM data,
- Adding attributes to CAD/CAM data,
- Enabling users to query data and attributes.

DCS/CDF also provides communications functions in order to transfer data between the repository and CAD/CAM applications.

Presentation in the CIM Environment

Presentation in the CIM environment means providing data to and accepting data from people and devices. Of course, this data must assume appropriate data definitions and screen formats to be usable.

Because today's industrial enterprise contains such a wide array of devices and information needs, it must have a consistent way to distribute and present information to people at terminals or workstations, machine tools, robots, sensors, bar code readers, automated guided vehicles, and parts storage and retrieval systems. The range of this information covers everything from simple messages between people to large data arrays for engineering design applications. It may originate from a CIM user in one functional area of the enterprise and be delivered to a CIM user or device in another area.

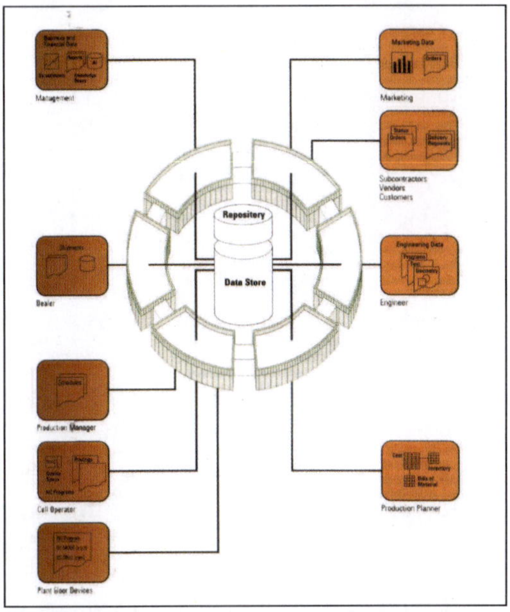

Presentation in the CIM Environment.

In today's environment, presentation occurs on displays that utilize different technologies. Some are non-programmable terminals, some are programmable workstations, and are uniquely implemented for each application. As a result, the same information is often treated differently by individual applications.

For example, the same manufactured part may be referred to as a part number in a bill of material application in Production Planning as a drawing in Engineering's CAD application and as a routing in a paperless shop order application from Plant Operations.

As data is shared across the enterprise, it must be transformed into definitions and formats that support the needs of each individual user and application. And applications must be able to access shared data, collect the required information, and then format that information for delivery.

Computer-aided Process Planning

A detailed process plan usually contains the route, machine and tool, processes, and process parameters. A basic CAPP model is shown in Fig. Since there are multiple choices for machining operations for the same machined surface, multiple machine tools available to perform the same operation, and different machining sequences, etc., alternative decisions can be made in process planning. For a machined part, there usually exist several feasible process plans for creating the part, although most existing CAPP systems are designed in such a way that only one process plan is generated. Usually, these CAPP systems work in conjunction with CAD and other computer systems. Sequential design and manufacturing follows a line path, with each step beginning after the previous step is completed.

In short, CAPP is a decision-making process. It determines a set of instructions and machining parameters required to manufacture a part. There are four main elements in designing a CAPP system: input, output, database, and manufacturing decision-making rules. As shown in the CAPP model, data preparation is a necessary step for obtaining proper product definitions for the CAPP system which is carried out after the CAD system is completed. After the completion of data preparation the data is entered into CAPP and is supported by knowledge and physical rules. When the CAPP operation is completed, the next step is output. From this output, the next operation is post processing which prepares data for production planning and scheduling activities. Finally, the production planning and scheduling operations are carried out. Those components in the dashed-line box are independent of CAD and production planning/scheduling.

There are generally two approaches to CAPP systems, namely variant and generative.

The variant approach was used in early computer-aided process-planning systems, and is basically a computerised database retrieval approach. The variant or retrieval approach is based on group technology methods of classifying and coding parts for the purpose of segregating these parts into family groups. In this approach, parts produced in a plant are grouped into part families, distinguished according to their manufacturing characteristics. For each part family, a standard process plan is established. The plan is stored in a computer file and then retrieved for new parts that belong to that family. Some form of parts classification and coding system is required to organise parts into families for correct retrieval of the appropriate plan for a new part. A major problem with this approach is the lack of adequate classification models that can provide consistency in classifying and coding parts. It is also restrictive in that new parts to be planned have to be similar to those already in the data file. The second approach to computer-aided process planning is the generative type. Systems of this type synthesise the process plan for a new part, based on an analysis of part geometry, material and other factors that may influence manufacturing decisions. Inputs to the system would usually include a comprehensive description of the part. This may also involve the use of some form of part coding, but this does not involve the retrieval of existing standard plans. These systems usually employ either a set of algorithms or knowledge-based techniques to progress through the various technical and logical decisions toward an appropriate process plan for a part. The generative approach provides fast advice to designers early in the design process and is closely coupled with the product-modelling activities. Once the manufacturing technology and the type of equipment or process have been chosen, further detailed planning is carried out as usual. The use of knowledge-based systems and artificial intelligence techniques was the next major development in generative process planning.

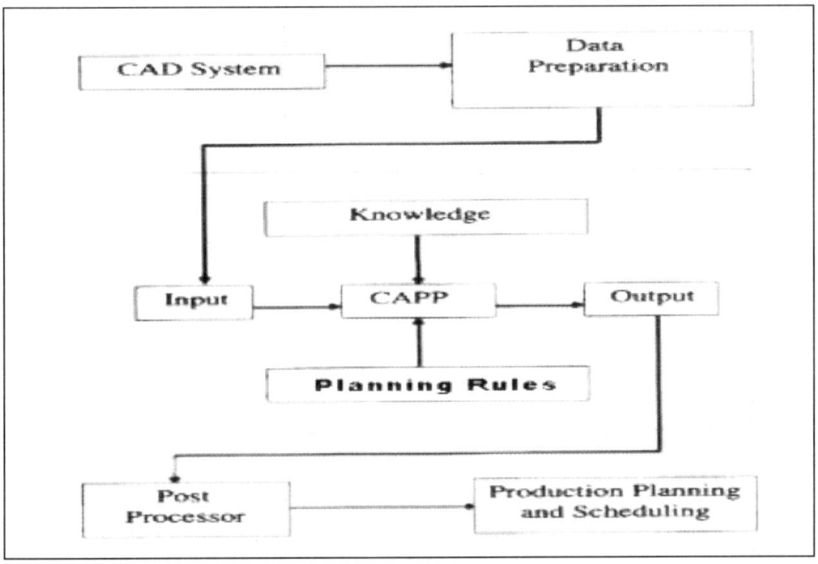

Basic CAPP model.

Need for Process Planning

This is the planning strategy for manufacturing the part. It is defined as the activity that translates part design specifications from an engineering drawing into the manufacturing operation instructions required to convert a part from a rough to a finished state. It represents the link between engineering design and shop-floor manufacturing. It is a major determinant of manufacturing cost and profitability. The gap between CAD and CAM can be shortened considerably by developing better systems for process planning.

There are two levels of process planning:

- High-level planning: The planner identifies the machine's features (surfaces) of the part, groups them into set-ups, and orders these set-ups. The features to be cut in each of the set-ups, and the tools for cutting each feature.
- Low-level planning: Specifying the details of performing each step that results from the first level such as choosing machines, cutting conditions (speed and feed), type of fixture, cost and time estimates, etc.

Approaches of Process Planning

There are three approaches of process planning:

- Manual approach,
- Computer assisted variant approach,
- Computer assisted generative approach.

The later two are known as Computer Aided Process Planning (CAPP).

The following are some of the drawbacks of manual approach:

- The approach becomes rapidly inefficient and unmanageable when the number of process plans and revision to those plans increase. Consequently, inconsistent plans and large time requirements for planning of ten result.
- The plans by this approach reflect the personal experiences and preferences of the process planner.
- This is considered a poor use of engineering skills because of the high clerical content in most of its functions.

These limitations have logically leaded to computerized approach to process planning.

Conventional Process Planning

Traditionally, skilled planners perform process planning. Planning is based on planner's decisions. The quality of the plan highly depends on individual skill, knowledge and experience. Planning is very time consuming, because of voluminous routine work and calculations. Only little time is left for creative work and updating planner's knowledge. The functions of process planning in the conventional manufacturing industry are shown in figure.

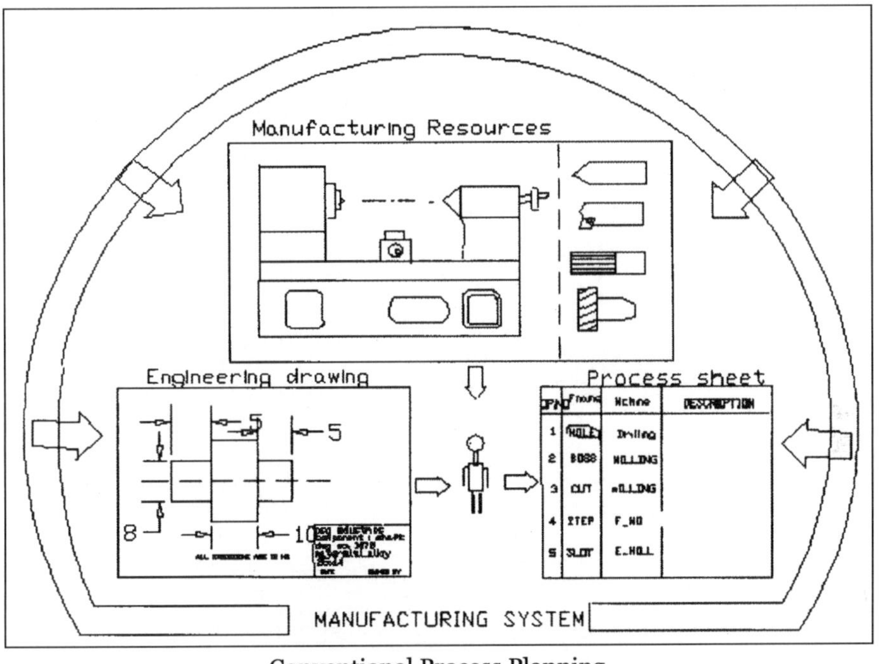

Conventional Process Planning.

Importance of Computer Aided Process Planning

The importance of CAPP has already been identified by industries. Many companies in manufacturing sector have started developing CAPP for their internal use. The aim of Computer Aided Process Planning is to eliminate or at least reduce the manual intervention in process planning. The role of Process Planning from design to part is shown in figure.

It is also found that there is generally a lack of consistency among process plans prepared by different individuals with varying manufacturing backgrounds and levels of skill. It is therefore imperative to automate the process planning function and this makes CAPP important even in the absence of CAD and CAM. By using a computer, the tedious and repetitive aspects of process planning can be speeded up and this helps to optimise the total manufacturing function by releasing the experienced planners and enabling them to concentrate on those aspects outside the scope of a computer. At the same time, more consistent process plans can be obtained by applying a standard set of rules, which increases confidence in the system and helps in the rationalisation of production. To automate process planning, the logic, judgement and experience required for process planning must be captured and incorporated into a computer program.

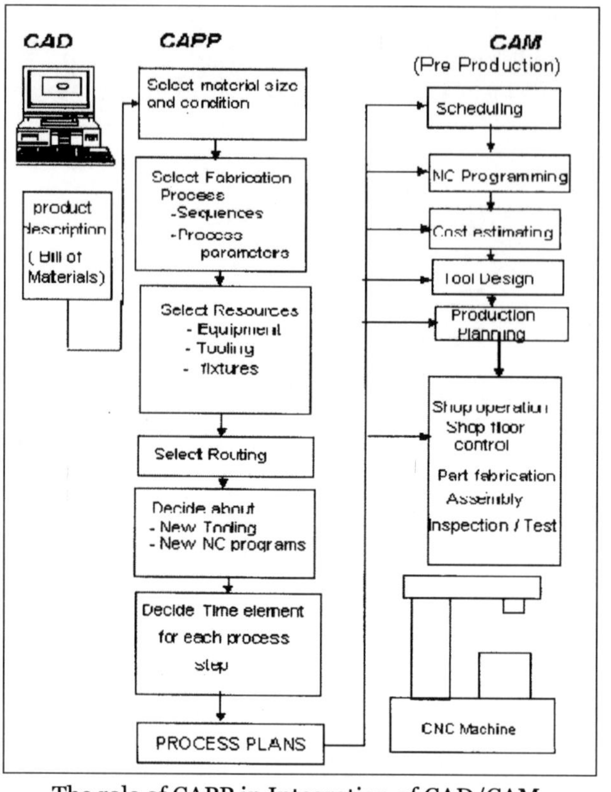

The role of CAPP in Integration of CAD/CAM.

Conventional Process Planning versus CAPP

To manually plan processes for complicated parts, the process planners must be quite experienced persons. A Process planner usually spends considerable time to plan a process and complete the process documents depending the complexity of the part. It is thus evident that the development of a CAPP for the prismatic components is even more meaningful. The aim of Computer Aided Process Planning is to eliminate or at least reduce the manual intervention in process planning.

The advantage of CAPP over conventional process planning is manifold:

- It improves the quality and consistency of process plans.
- Savings in lead time: In today's concepts of TQM and batch manufacturing lead time seriously effects the delivery time of the product which plays a major role in deciding the market share of any company.
- Process rationalisation: CAPP generated process plan is more likely to be consistent (because the use of same software for all plants), logical and optimal (because the company has presumably incorporated the experience and judgement of its best manufacturing people into process planning software and computer can evaluate more number of alternatives in a less time).

- Integration with manufacturing support functions: These functions involve NC part program generation, computerised work standards etc., if done independently these functions involve duplication of clerical work.
- Increases the efficiency of process planning.
- Decreases the variety of tooling.
- Error reduction: Helps in overcoming the casual errors committed by human process planners. If the error goes unnoticed, it penalises heavily from the cost point of view.
- It can reduce the skill required by planners.
- It can increase productivity.
- It can consider all available equipment while process planning.

Nobel first discussed the use of computers in process planning. Early investigation of automated process planning was done at Purdue University by Scheck and by Berra and Barash. Many investigations were updated in the late seventies, and research work gained momentum in eighties. CAPP is the application of computers to assist the human process planner in process planning function. In its lowest form it will reduce the time and effort required for preparing process plans and providing more consistent process plan. An advanced CAPP will provide the automated interface between CAD and CAM.

Automated process planning makes use of CAD model as input to the system. The subject of features and their role in product design, process planning and many other activities spanning the product life cycle is playing a vital role in today's technology. This method extracts simple polyhedral features like slot, step, etc. But it fails to extracts V slots, T slots, curved features and complex features.

One of the widely used CAD/CAM packages, CADDS5, has been used in the development of this package. Even if the component is modeled in other modeling packages, we can use this software by adding a simple routine which will convert the part internal representation in that package into linked lists of Faces, Edges and Vertices whose structures are defined in the software.

Emergence of CAPP

Design, planning, manufacturing, assembly, inspection and testing are major stages of product development. Computer based technologies are available today to carry out most of these activities except process planning. Process planning is a heuristic and knowledge based activity and is one of the missing elements in the integration of CAD and CAM. Though process planning covers a wide spectrum of manufacturing operations like machining of components, sheet metal processing, and other secondary manufacturing operations, machining offers considerable challenge for computerization. Flere too, process planning of prismatic component is more involved than rotational component. Processing of prismatic components requires the use of several machines like precision milling, boring, drilling and tapping and these operations can also be performed in a CNC machining center. Proper sequencing of operations, selection of right cutting conditions and tools, proper filtering etc. are necessary for satisfactory realization of the components, meeting the high level of precision, and accuracy required for application in missiles. Many researchers have attempted this problem.

Process Planning establishes the methods and means of converting the raw material into a finished part. The planning function interacts with the manufacturing system through:

- The design specifications,
- Manufacturing resources,
- The process plan.

Its basic task is to determine by what means and how a product is to be manufactured economically and competitively.

Process planning involves, according to the part specification, determination of processes, machine tools, cutting tools, operation sequences, machinability data and calculations of time and costs. The results obtained must be reflected in documentation.

Computer aided process planning should diminish as much as possible deficiencies in traditional process planning.

A set of process plans would include:

- Design input,
- Material selection,
- Process sequence,
- Machine and tool selection,
- Fixture selection,
- Machining parameter selection,
- Cost/time estimation.

Clear-cut pictures about the necessities have to be formed before starting the development of a project. George P. Sutton had done a survey for CAM-I (Computer Aided Manufacturing - International) on CAPP practices, plans and future needs in 1987.

Fifty companies participated in this survey and it was found that there were no fully automated CAPP systems and almost all of them required extensive manual operations. However benefits that accrued to those companies that had a semi manual CAPP system were substantial. The survey indicated a user desire for standardizing the interfaces of CAD with CAM, as well as with other manufacturing oriented systems (like scheduling, inventory control, cost estimating, etc.). There seems to be agreement that CAPP is an essential element of future CIM systems.

The first CAPP system was developed under the direction and sponsorship of CAM-I (Computer aided manufacturing - International) and was presented at the "1976 NC conference". In the same year MIPLAN was developed by the OIR (Organization of Industrial research) since then CAPP has begun to be widely addressed. Meanwhile, due to the increase in competition, companies started implementing integrated manufacturing. CAPP as a main element in the integration of design and manufacturing has not kept pace with the development of CAD and CAM. This situation made the process planning, a bottleneck in the implementation of integrated manufacturing. Thus more and more effort has been applied in the CAPP area and numerous CAPP systems have been reported.

CAPP Systems

ASCAPP

ASCAPP is a generative CAPP system for process planning of aircraft structural parts. This has been developed at Nanjing University of Aeronautics and Astronautics, Peoples Republic of China. In the developed CAPP system a 6- orientation, multi-level feature oriented method is adopted for part description. The technological decision module of the ASCAPP system is an expert system that is composed of a knowledge base, a inference engine and a data base for storing part information and interim decision results.

APSS (Automated Process Selection and Sequencing)

In APSS, the 'bottom-up' or backward planning approaches have been adopted. APSS has been written in PROLOG and has been implemented in the prototype knowledge based CAPP systems, ACES (a CAPP expert system), which is an automated process planning system for the manufacture of complex prismatic machined parts.

OPPS-ROT

OPPS-ROT is generative CAPP system for rotational parts. This has been developed at university of Gazintep, Turkey. The sequential operational planning method is used and it is based on the assumption that a single feature on a component usually considered as a single operation. It is implemented with C++ on a personal computer.

APPAS, CADCAM and TIPPS

APPAS is one of the first generative CAPP systems. Wysk developed it in FORTRAN. Process planning logic has been implemented using the decision tree approach. The input to the system is text input in the form of special codes. CADCAM is an extension of APPAS, developed by Chang and Wysk, which links APPAS with interactive computer graphics terminal. TIPPS is new generation of APPAS and CADCAM, once again developed by Chang and Wysk. It is one of the first systems that integrate CAD and generative process planning into unified system utilizing the AI and decision tree approaches.

AUTAP and AUTAP-NC

AUTAP is one of the complete generative process planning systems developed in the early stages. AUTAP is capable of determination of raw material, operation sequence, selection of machine, calculation of estimated time.

CMPP

CMPP is a generative system developed in FORTRAN 77, aimed at high technology machined cylindrical parts. Parts characterized by expensive material, tight tolerances and complex machining processes are most suitable for this system. CMPP is interfaced to many CAD/CAM systems in American aircraft companies.

EXCAP and ICAPP

EXCAP is an expert system aimed at machining of rotational parts. Rule based technique with backward planning mechanism is implemented in this system. ICAPP is a variant system developed in FORTRAN for prismatic parts. ICAPP is feature oriented interactive system, capable of eight basic machining processes.

GENPLAN

GENPLAN is generative system developed in 1981. It uses a GT based coding system that covers part geometry and process variables to generate comprehensive operations sequence. GENPLAN can be interfaced with management decision support systems, automatic shop loading and scheduling, tool order processing systems.

GIFTS

GIFTS is a generative interactive feature-based system for process planning of rotational components. This was developed at IIT Delhi. GIFTS selects the operations, machines, tools, cutting parameters for each feature by taking the relevant data from the data base and finally sequences the operation.

Part of the slow progress in this area is due to the complex and dynamic nature of the planning domain which possess a great challenge to the research community. Because our present understanding of process planning activities is limited to local micros, we can attain some success in programming specific kinds of process planning activities (e.g., to machine a cylindrical part), but are unable to automate the whole process which would require a much broader understanding of the necessary tasks. Currently, the variant and semi-generative approaches seem to be the most practical ones, and these systems are readily amenable to certain real-world applications.

Objectives of Process Planning Systems

Notable requirements for process planning systems are consistency, accuracy, and ease of application and completeness.

Consistency

One of the most common problems of manual process planning is the existence of inconsistent process plans. Consistency is essential not only in the application of data and knowledge to generate process plan, but also in the acquisition and presentation of manufacturing information. Process plans prepared by different process planners will vary depending on the variations in their knowledge, process logic and experience etc., a good process planning system should provide consistently same process plan for same set of input data.

Accuracy

Accuracy is another important objective to maintain in any process planning system. Inaccuracies in process plans and corresponding information in them can greatly undermine the integrity of the system.

Ease of Application

This is another important trait, which determines how quickly a response can be made to generate a new process plan or make revision to an existing process plan. This is of particular importance in today's short lead-time environment.

Completeness

Information about manufacturing processes, machines, tools, cutting parameters etc., must be completely known while decisions about process planning g are made. If complete data is not provided, then the choice can easily lead to inconsistency in processing methods and corresponding inconsistencies in part quality and costs.

Challenges in the Area of CAPP

The complexity of manufacturing activities makes it impossible to generate an ideal plan in the first time. In practice, it usually takes much iteration based on feedback from the plan execution stage to gradually refine a plan into its optimal form. Most current planning activities are mainly concerned with the plan generation phase; no major effort has been made to link plan generation with its execution and monitoring activities. But since the execution of a manufacturing plan normally spans a period of time with the involvement of multiple departments, this manual monitoring and feedback task is difficult to accomplish effectively in practice. The lack of this self-improving ability due to the open-loop nature is one of the reasons why most CAPP systems are not up to the mark.

Current research efforts tend to focus on a specific kind of planning function for a specific type of part. These micro-view point planning approaches produce computer systems that perform individual tasks in isolation from other planning activities. They result in many small, and still isolated, islands between the islands of design and manufacturing.

Due to this micro-view point of planning activities and the lack of an overall planning structure, most planning systems developed to date focus on a narrow range of activities, which severely limits their applicability in practice. For example, most efforts in current generative CAPP systems are aimed at generating machining sequences of certain limited part shapes. Since they are based on machining handbooks and databases, decisions made at the operation-planning level are treated as a black box, without detailing how each sequence will be carried out in actual operation.

There is no single best approach that can handle all the functions required in an integrated planning system. Hybrid approaches, which logically combine the strength from several different techniques, should be devised and used. For example, various techniques from optimization and operation research should be incorporated into plan sequence generation and mathematical models of processes should be part of the operation planning systems.

Process planning systems, in order to provide integration, automation and flexibility should:

- Be generative,
- Be technology based,
- Use features as a technological and communicational interface between design and process planning,
- Be able to extract automatically all product data,
- Use a supervisory control system to ensure user-friendliness and flexibility in use,
- Integrally support all planning tasks, including capacity planning and scheduling,

- Take decisions based on optimization techniques,
- Be fit for close-loop planning.

Approaches to CAPP

Traditionally two approaches are recognized for computer aided process planning, the variant approach and generative approach. In variant approach the GT principles are used. Further making use of the similarity in design or manufacturing of the parts are used in producing the new process plans. In generative CAPP the process plan is generated automatically, the input being the details from CAD drawing. However with the rapid development of new techniques, many CAPP systems do not exactly fit into this classification and combines both approaches.

Variant Approach

The variant approach is comparable with the traditional manual approach where a process plan for a new part is created by recalling, identifying, and retrieving an existing process plan for similar part, and making necessary modification making it suitable for the new part. In general this system has two operational stages.

- Preparatory stage: In this stage the parts are coded and families are formed based on group technology (GT) principles. Standard process plans are prepared for each family and stored in database.
- Production stage: Process plan for a new part can be made in this stage. Whenever a new part comes for process planning, its code and family will be identified and the standard processing plane for that family is retrieved. Necessary modifications are made by the process planner to satisfy the component design.

The CAPP system may be either semiautomatic or automatic. In the semiautomatic there is a need for manual input of the required data. Inputting the feature details, finished material details and raw material details to the computer is time consuming and tedious. This is a form of underutilization of man power and here are also chances that the user may input the data wrongly. These drawbacks can be avoided by providing a bridge between the generated automatically from CAD model or drawings. This will also help in enhancing the role of computers in industry. This helps to move towards computer-integrated manufacturing whose is a paperless industry.

The majority of existing CAPP systems is of variant approach. Some of the processes planning systems that follow this approach are CAM-I CAPP, MIPLAN, MITURN, MIAPP, and ACUDATA. The disadvantages of this approach are that the quality of process plan still depends on the knowledge of process planner because of editing. Process planning for an entirely new type of component is not possible.

The Generative Approach

Generative process planning synthesizes process information in order to create process plans for a new component automatically. The input of such systems may be either CAD drawing or models or text input. Knowledge of process planning is captured and implemented in software using different techniques. The process logic can be implemented in the system by means of decision tables, decision trees, axiomatic, rule based or constraint based. The decision logic consists of checking for some conditional requirements of the component and selects the processes. Other planning functions like machine selection, tool selection, process parameters selection etc., can also be automated using the generative process planning techniques.

The biggest advantage of this approach is that process plan developed is consistent, and fully automated. This is especially useful for product mix of number of products with small lot sizes. APPAS, TIPPS, AUTAP, GENPLAN, TURBO-CAPP, TOM and ICAPP are some of the generative process planning systems.

Semi-generative Approach

This approach is the combination of variant and generative approaches. However, compared with the variant systems. The advantage of using such systems is reduced time for preparation of process plans; modifications of generated process plans are necessary in order to implement in the real work.

Implementation Techniques

Group Technology

The typical utilization of GT is in the part family concept where coding and classifying of the part are done. The GT principles are mainly used in the variant process planning systems. CAPP-I, SAPT, TOJICAP, WICAPP are some of the systems which are implemented using the GT principles.

The Bottom-up Approach

This approach develops the CAPP systems by filling in materials to convert a finished part back to the initial blank.

This is similar to conventional computer assisted process planning method oriented towards the variant approach.

The Top-down Approach

This approach develops the CAPP system by means of tracing the task of process planning from top to bottom i.e. from raw material to finished part. This is an automated computer aided process planning method oriented towards the generative approach. This requires the general rules of manufacturing strategy to be built into algorithms which can operate on brief input data describing the geometric features and engineering requirements of part.

AI and Expert System Techniques

An expert system can be defined as a tool that has the capability to understand the problem specific knowledge and use the domain knowledge intelligently to suggest alternative paths of action. CAPP systems using this technique are called as either expert system (ES) or knowledge based system (KBS). The decision logic is implemented using the IF and THEN rules are frame-based rules. TOM, GARI, EXCAP, KAPPS, XPLAN are some of the CAPP systems in which these techniques are implemented.

COMPLAN System

COMPLAN is a generative/variant of CAPP systems. This process-planning system has been designed for small batch manufacturing of mechanical parts and is mainly developed in the C++ language. The modules of the COMPLAN system can be arranged in two main functional groups: process planning and workshop scheduling. The relational database is central to the system and is used to exchange and synchronise information between the modules. The COMPLAN architecture provides a framework in which all functional modules work together in a concurrent approach. The COMPLAN system follows a two-level hierarchical approach to scheduling which is performed in different time horizons, levels of aggregation to achieve better load balancing, and to obtain accurate resource assignments and rescheduling functions.

Economic competition is ever increasing, especially since the globalisation of the manufacturing business. In order to anticipate new challengers, manufacturers are urged to revise their economic objectives fundamentally. The COMPLAN system realises economic improvements such as:

- A shortening of the lead time from design to manufacturing and reducing the effort spent to enter CAD data into a CAPP system, and by rationalising the process-planning effort needed.
- Achieving a leaner and more flexible production organisation.

The most important feature of the COMPLAN process-planning system is its ability to handle nonlinear (manufacturing alternatives) process plans. The system provides either manual or automatic planning functions for process planning and scheduling. So, the user will at all times retain full control over the system-planning output.

ESTPAR System

ESTimator of PARameters (ESTPAR) is a generative approach to a CAPP system. The knowledge-based expert system along with the GP code is called ESTPAR. The GP code was written in FORTRAN IV, and ESTPAR has been written in FORTRAN 77 for the convenience of integration with the GP code and the macro-level approach. The ESTPAR system can be used by manufacturing personnel to determine the optimal machining parameters and the corresponding machining costs when using different combinations of machines, tools, and fixtures (MTF). The ESTPAR system is also a part of a large automated process-planning system to be used by design and manufacturing engineers for efficient part design and manufacturing. The main advantage of this system is that more machining operations can be included and more functions of automated process planning can be added without any difficulty.

FCAPP/SM System

FCAPP/SM is a generative CAPP system. This process planning system, which has been developed for sheet metal parts, has incorporated several features of generative systems using a procedural decision making framework. The system was developed using the Clipper language compiler. Clipper works with dBASE-compatible databases and allows the development of stand-alone systems. The part storage system consists of two main databases and several support databases. The main parts database contains one record for each part. This record stores the part name, raw material form, material, overall dimensions and tolerances, and other part-specific information. The feature database contains one record for each feature of a part. Features currently identified are holes, slots, and bends. Hence, a many-to-one relationship exists between the feature database and the parts database. The databases are linked by the part number field common to both databases. The FCAPP/SM system contains about 30 rules which match the part attributes from the part databases with the standard machine information found in a machine database. The process planning strategy adopted in this system is to arrange all the features of the part into a predetermined precedence hierarchy. Corresponding to each feature, all feasible machines are identified next. This system has the capability of adjusting to real time problems such as machine breakdown, job priority and overloading. Unlike existing systems which use assumed weights and heuristic search strategies to obtain a feasible plan, this system uses cost data and optimisation methods to generate an optimal plan.

IKOOPP System

Intelligent knowledge-based objective-oriented process planning (IKOOPP) is a generative CAPP system. Because of the problems associated with manual process planning, and the inconsistencies of the plans, the IKOOPP system has been developed to automate and standardise the process planning function for the manufacture of progressive dies. The IKOOPP system receives part definition data from a commercially available die design system called Auto-Trol Series 7000 Die Design. A feature extractor has been developed to extract the relevant attributes of machining features within a progressive die plate mode. These attributes are used by the IKOOPP system to reconstruct the plate model using an object-oriented representation for the subsequent reasoning processes. Knowledge of the functions of the machining features is used to deduce engineering information which cannot be represented easily in the CAD system. Moreover, the available interface for accessing the Auto-Trol databases is limited to a set of FORTRAN subroutines. The IKOOPP system automatically plans the set-up sequences, and selects the required machine tools, cutting tools, heat treatment, fixturing elements and sequence of operations.

KAPLAN System

Knowledge-based approach to process planning (KAPLAN) is a generative CAPP system. KAPLAN provides fully automatic generation of productions plans of rotational parts. The program structure is based on knowledge base techniques and the knowledge required for the plan generation is represented by IF-THEN rules, easily adapted to every workshop environment by means of a user interface. The application of this technique is particularly effective where flexible plants and machine tool are used for a widely variable group of products, e.g. in flexible manufacturing systems. The module for the automatic generation of turning sequences, (KAPLAN) is a part of a more complex process-planning system for the programming and the control of a flexible cell. A great advantage in terms of time and data reliability" is obtained by this CADCAPP integration. The technique for generating plans used by this system is quite different from those used by the other systems currently proposed and represents a new approach in CAPP systems for rotational parts.

Table: Details of the CAPP systems.

System	Approach	Programming language
AFR	Generative	
ALPS	Generative	Clipper language compilar with dBASE
AMOPPS	Generative	Turbo Pascal
CADEXCAP	Generative and variant	Fortran 77 and POP 11
CAMSS	Variant	
COMPLAN	Generative and variant	C++
CROPS	Generative	OPSS
ESTPAR	Generative	Fortran 77
FBD	Generative	
FCAPP/SM	Generative	Clipper language compilar with dBASE
GEOPDE	Generative	Prolog
GFAS	Generative	ECSL
GLM	Variant	

IKOOPP	Generative	Fortran Auto-Trol series 7000
KAPLAN	Generative	IF-THEN 1,ale
K-base	Variant	Turbo Pascal
MCOES	Generative	LISP
PART	Generative	
PerMIA	Variant	
RDCAPP	Generative and variant	FORTRAN SQL
ROBEX and RATE	Generative	OPSS
SMT	Generative and variant	LISP
TAB	Generative and variant	LISP
TAMCAM	Semi generative	dBASE
TVCAPP	Generative	Turbo PascaL

K-Base System

A knowledge-based (K-B) system is a generative CAPP system. Initially this system was written in Prolog, but when it was redesigned for more generic applications, it was re-coded in Pascal, and runs on IBM compatible computers. A knowledge-based system is used for process planning and cost estimation in the hole-making process. The main function of the system, besides estimating the cost of production, is to recommend appropriate processes, their sequence and their respective machining conditions in order to obtain the required product specifications. The knowledge required for process planning and cost estimation is organised into three knowledge bases, namely process and sequence knowledge base, machinability data knowledge base and costing data knowledge base. In comparison with the manual system previously used, this system has provided several advantages including the flexibility to change data, and uniform process plans, correct machining parameters, and automatic cost estimation. Another important feature of the system is that it provides a company with a facility to store the knowledge gained by experienced planners in the databases which can then be used to assist inexperienced planners to perform the task of planning and estimating quickly and efficiently. A major feature of this system is that it unifies process sequence, machinability, and cost estimation into an integrated system which caters for the requirements of small to medium sized companies involved in batch operations.

MCOES System

Manufacturing cell operator's expert system (MCOES) is a generative CAPP system. The architecture of the MCOES system consists of five main subsystems:

- A design data interface.
- A process plan preparation system (strategic process planning).

- A process plan generation system (operative process planning).
- A method editing interface.
- A factory modeller.

The overall operation of the system takes place in three stages:

- System set-up stage: When a factory model representing factory facilities is generated and a collection of methods representing tested, proven manufacturing processes is created. In this stage, the planning focus is on a single process.
- Strategic process planning stage: When a process plan specification for a new product family is created by choosing appropriate methods and set-ups. In this stage, the planning focus is a single product family, comprising several processes.
- Operative process planning stage: When a detailed process plan is created for a given manufacturing order. In this stage, the focus is on a single order, possibly comprising several instances of several product families.

The main advantage of this system is that it is easy to generate correct part programs on the basis of relatively high-level order information, in a few minutes.

ROBEX and RATE

RObot Based EXpert (ROBEX) system, and robotic assembly time estimators (RATE) are generative CAPP systems. Robot-based flexible assembly systems are complex and their planning and development requires a methodological planning procedure. The two systems are designed as a part of an overall planning procedure for the design of robot-based flexible assembly systems. This seven-step planning procedure provides an integrated approach to robot integration in CIM systems, from the initial feasibility studies to the final detailing of the flexible assembly system design. ROBEX and RATE belong to the second step, the determination of basic data. The purpose of this step is to determine the information which will be needed in the subsequent layout and planning of the assembly systems, i.e. a feasible set of assembly operations by which the product can be assembled, and estimated assembly times can be determined for this set of operations. Highly accurate assembly times are not necessary at this step in the planning procedure, what is required, and what RATE provides, are estimates of the operation times, which can be used in the subsequent simulation and evaluation of the proposed assembly system. The lead time between design and manufacture of production can easily be reduced when using these systems.

TAMCAM System

TAMCAM is an open manufacturing system; developed by Smith et al. Three-level hierarchical control architecture is employed in TAMCAM. Based on the functional characterisation of shop floor control functions, a set of procedures for testing the intelligent control of a manufacturing system has been set up. In TAMCAM, part design and process planning is integrated in the AutoCAD (Autodesk, Inc.) environment on a Unix platform. Each machinable feature of a part is drawn in a dedicated AutoCAD layer. The process planning system used in TAMCAM adopts a semi-generative approach, i.e. the standard process plans for a set of part families are stored in a database, and changes in a part design are resolved using decision-tree logic. The standard process plan for a part includes operations, resources, machining parameters, machine independent cutter location data (CLDATA) for each feature, and machining precedence of the features. The process-planning system can generate new process plans (including CLDATA) for a feature when the feature is changed in terms of location, shape, and tolerance. The advantage of this system is that a number of open system standards for CIM have been developed to reduce the costs associated with software development, integration, and maintenance.

TVCAPP System

Tolerance verification in computer aided process planning (TVCAPP) is a generative CAPP system. The TVCAPP is an expert system, and it has 3 components:

- Databases.
- Qualifiers and variables.
- Algorithms and external programs.

TVCAPP integrates the combined effects of process selection, machine selection, tool selection, machining parameters and tolerance requirements in developing more efficient alternative process plans. The main advantages of the TVCAPP system over existing ones are that it:

- Verifies the input values of tolerance by the user or from an AUTOCAD drawing.
- Checks tolerance values against standards (CSA-B78.2-86).
- Generates four types of geometric tolerances: straightness, circularity, cylindricity and concentricity.
- Generates a limited number of operations for the statistical process control analysis.
- Generates alternative process plans in which each operation has an operation index.
- Calculates the cost for each process.

The proposed expert system should lead to the development of a systematic means for automatic CAD/CAPP integration from a given CAD drawing, machine and tool databases.

Computer-aided Quality Assurance

Computer-aided quality assurance is when computers are used to inspect and test products that are being manufactured. It is an engineering application that can oversee the operations of machinery and ensure everything is being produced as required.

It allows for a digital inspection of the quality of the products being produced. This includes measuring the equipment and its management/handling, producing a vendor rating, conducting a goods inward inspection, creating an attribute chart, putting in place statistical process control (SPC) and producing necessary documentation.

Computer-aided quality assurance can also ensure a failure mode and effects analysis, otherwise known as FMEA, is put into action. Additionally, it can allow for advanced product quality planning and a dimensional tolerance stack-up analysis on computer-aided design models using the product and manufacturing information.

On top of this, it also encompasses a computer-aided inspection with coordinate-measuring machines and can compare data that has been gathered through the 3D scanning of the physical elements of computer-aided design (CAD) models.

Application of Computer-aided Quality Control System

There are many reasons that make the quality of products poor. The main reasons are as follows:

Indifferent Quality Consciousness

Due to the long-term influence of the old economic system, some enterprise managers are only interested in the quantity and speed of the production, and the workers care only how to finish the production task and product quantity, so there is little enthusiasm to improve the quality. "Quality the first and Consumer the supreme" is not implemented in the work. Generally speaking, most enterprises have not established a sound quality culture to guarantee the quality to be raised with a long-term and steady step.

Serious Formalism

Depending on the government administrative means, the Total Quality Control is carried out under the old economic system. Because there is no strong sense of quality crisis and no enthusiasm to raise the quality, the formalism has taken place and spread. The theory and methods of quality control has become the showing vase. Even the quality training has become the victim of the Formalism, which is usually used as the capable method of strengthening the workers' quality consciousness and raising the quality management ability, and it makes the indifferent quality consciousness even more superficial.

Poor Work Quality

Because of the indifferent quality consciousness and existence of Formalism in quality training, some basic quality management methods and some simple and practical quality management methods appeared recently have not been popularized, such as Local Control Chart, Chain Technology. Therefore our workers are usually shortage of methods and techniques of controlling and improving the quality.

Many workers have a little knowledge in the most basic methods of quality management such as Pareto Chart, let alone monitor the working process with the Control Chart. These three aspects tie up and influence each other; the indifferent quality consciousness will certainly foster the Formalism in quality consciousness. These two elements again constrain the reform and development of management methods. That the backwardness of quality control methods causes the low effective quality management will similarly encourage the Formalism and limit the quality consciousness.

Under such a vicious circle, how could the enterprises extricate themselves from the difficult position, how could the enterprises improve the quality of their products?

Modernization of the Control Methods-breakthrough to Improve Quality

If China wants to get rid of predicament of quality control, the only way is to strengthen the nation's quality consciousness, foster the perfect quality culture and completely eradicate formalism. This is a basic guarantee for an enterprise, even the whole country to gradually raise the quality of products. However, it is not impossible to raise the quality of products by the large margin because the improvement of control methods will encourage the improvement of the quality. It will be very effective for using computers to control the key processes.

The Necessity of Quality Control with Computers

The Need of Guaranteeing the Normal Production

Under the market economy, the quality is the basic guarantee for the enterprises to exist and develop. In the intense competitive market, the enterprises more and more realize the importance of quality of products. The fact is that good quality of products is accomplished by the corresponding quality control methods. For example, there are 5,000 workers in American Kodak Color Film Company, but 35,000 control charts are used, the average number is 7 per worker. From this, we can know how much the high quality products will rely on the advanced quality control methods.

Using computers to control the quality will be a direct way of encouraging the enterprises to get rid of the vicious circle of the quality control. Then the operators' tasks of quality control which are usually out of their abilities can be shared by computers to a great extent. The influence of the workers' professional performance and level upon the quality can be reduced, the quality and the quality control level can be improved, and then the economic performances can be increased, the enterprises can invest more to offer the quality training to the workers, and the problem of modernizing the quality control methods can be completely solved.

Going with the Trend of the World Quality Control Development

When the quality control work becomes more and more complicated, the quality data inspected manually by the quality inspectors and operators are far from meeting the need of modernized production. The application of the computers makes the quality control more effective and standardized. At present, computer aided quality control plays very important role in some developed countries, while in China it is still under research. The great potential of the quality control with computers has not been developed. So using computers to control quality can not only follow the trend of international quality control development, but also bring along the quality control development in the enterprises, and encourage uninterrupted improvement of the quality. So it can be said that quality control with computers should be a right choice for the enterprises under such kind of environment, it has both good economic and social performances.

Feasible Methods of Quality Control with the Applications of Computers

There are various kinds of ways for the computers to be used in the quality control. Those can be generalized as follows: Firstly, automatic control of the single equipment unit. This method means that the equipment and computers are installed together. So it is very expensive, which only a few enterprises are capable of applying this kind of equipment into every key process. Secondly, network control of the whole production line. There is its own terminal with every process, and the data transmission is operated very quickly and conveniently, but the invest is too expensive. So it is only applicable in some modern enterprises. Thirdly, quality control with computers under CIMS condition. At present, the research of CIMS just starts. Besides, the cost is too expensive, so it is not the time yet to use CIMS. Fourthly, centralized control with computers in the key process. Every(several) production lines is/are equipped with a computer, and one quality inspector is responsible of collecting and operating the quality number of each key process and feed backing the results in time to the certain process to decide what measures should be taken.

With the four methods mentioned above, the fourth one is obviously applicable to most of the enterprises. Firstly, this method costs less, at the present price, one 386 computer with a printer cost less than 10,000 yuan Renminbi, most of the enterprises can afford to it. In addition, most of the enterprises have computers, and many of them have not made good use of them, some lie idle already, and some used only as the office equipment. So this method can also make good use of the old equipment that enterprises already have. Secondly, the quality inspectors in the enterprises are usually well educated, and are familiar with the theory and techniques of quality control, they can collect the data quickly, accurately and authentically, so the quality control can be accurate and effective; another reason is that with this method the investment returns quickly, suppose the output value of every production line is 3 million yuan, if the losses of the defect products decrease by 1% every year, then 30,000 yuan expenses can be saved.

Problems Facing by the Centralized Control of the Key Processes with Computers

Because of the quality control with computers is just at the beginning stage, at present there is no such a successful and practical software package of controlling and analyzing quality. In addition, there is shortage of computer programmers of this kind. This greatly restricts the widespread application of controlling quality with computers. So in order to solve this problem, it is high time that a simple and easy operating quality control system was developed.

Function Design of Quality Control and Analysis System in Production

Before the software is designed and developed, at first the function requirements of the system should be fully investigated. Many enterprises hope that this quality control software should at least have the following functions: a) Function of analyzing the quality situation; b) Function of Quality control and alarming in production. c) Economic analysis of quality level. These function requirements of the system can be met through the methods described below.

Analysis of Quality Situation

There are many ways of analyzing the quality of products, such as Pareto Chart, Frequency Distribution Chart, while different methods have different emphasizes. With this system, the quality inspector should analyze quality situation of whole production and every key process, process ability situation between the preceding and next processes and the quality responsibilities taken by its process, so different kinds of methods should be adapted. The Pareto Chart can be used to comprehensively analyze the main quality problems of the production line and every process, then the direction of controlling and reforming the process can be pointed out; The chart can be used to judge the distribution situation of quality and to find out causes of these quality problems; The Control Chart can be used to analyze the situation of the process being controlled; The quality situation of the preceding and next process, to distinguish the quality responsibilities between the processes and to find out the emphasis of reforming. In order to make the system have the comprehensive analysis ability, the analysis module of quality situation should include Pareto Chart, Frequency Distribution Chart, Control Chart and quantitative analysis of two process ability indexes.

Quality Control Function in Processing

SPC plays a very important role in production. In SPC technology Control Charts are most often used. For ensuring the normal operation of the production, find out and remove the unusual, the quality inspectors need to use the Control Chart to analyze the situation of process being controlled. In order to meet different kinds of requirements from the production, the control module of production quality needs to have all kinds of control chart analysis functions. When the preceding process has influence upon the next one, the popular control chart is used to analyze the process control; when there is no influence, the Selecting Control Chart can be used to know every process's control situation.

Economic Analysis of Conformability Quality Level

How to appraise the economy of the conformability quality level of the process is also an important point in the quality control. It is not economical whether the conformability level of quality is too high or too low, the conforming quality can be shown by the process ability index, the conformability quality level corresponding to the best process ability indexes is the most economical one. Because there is a clear function between the process ability index and the losses of imperfect products, and there is also a relation between the process ability index and the investment that keeps the process ability to a certain level, then the best process ability can be determined by balancing the total expenses, so the basis are offered to appraise the process's quality level.

Importance of Computer-aided Quality Assurance

Quality control helps to mitigate risk, ensuring that consumers will not be receiving faulty goods that could be potentially dangerous. It also helps to protect the manufacturer in the event of this happening, and helps protect their liability when all correct procedures have been enforced.

- Save time: Using computer-aided quality assurance helps to save on development and manufacturing time, which therefore helps to reduce costs and potential delays. Products no longer need to be taken away over a period of time for manual sampling and testing, which can be time-consuming and has room for potential error.

- Streamlining: The quality control testing is also integrated with the manufacturing process, meaning they occur concurrently to one another. This means there is no need for separate quality control teams and saves on human resources. Instead, it can all be found happening simultaneously and automatically on the production line.

- Economical: The product is tested immediately upon being made without needing to be transferred between locations; this saves on travelling time for each batch of items and the potential for items to get damaged in the process. It also reduces the time it takes to make each item as the manufacturer doesn't have to wait for everything to be signed off by quality control.

- 100% perfection: Additionally, CAQC means that a 100% inspection can be conducted on every item being manufactured without too much difficulty. In traditional quality control testing, it isn't possible to test every one of the hundreds or thousands of products being made so only a sample can be taken.

  With computer-aided quality assurance, however, it is possible to run an inspection and testing on all products. This means you can absolutely guarantee the perfection of all items being manufactured and there is no risk or room for error.

  Also adding to this perfection is the fact that CAQC relies on non-contact sensors. This means that the item doesn't have to be handled manually in order to be inspected. As well as preventing the product from coming into contact with potential damage through physical handling, it also adds to the speed with which it can be conducted. The sensors can review the whole product in a matter of seconds and can be used along the production line.

- Data harvesting: The data collected from these sensors can also be used and interpreted to offer an analysis of the manufacturing process and identify any problems that require solutions. This will help to prevent further issues arising and will also mean processes can become more efficient and effective. If there are issues along the line, employees at the manufacturers will be able to easily identify where it is happening, what is causing it and what needs fixing.

Advantages of Computer-aided Quality Control

- 100% testing and inspection: In the traditional manual process the testing and inspection is done by the sampling process out of the hundreds and thousands of products or parts manufactured by the company since it is not feasible to check each and every product. With CAI and CAT hundred percent inspection and testing can be accomplished without much difficulty. With 100% inspection the company does not have to depend on statistical quality control method in which it is assumed that anything less than 100% of quality is acceptable. With computer controlled inspection, it is not necessary for the quality control department to settle for less than perfection.

- Inspection integrated with manufacturing process: In the traditional process there is separate quality control department where the manufactured product is taken for the inspection and testing. In CAQC the inspection process is integrated with the manufacturing process and it is located along the production line. Thus as soon as the product is manufactured it is tested immediately by the computerized process without moving it to some other location. This helps in reducing the overall time required for manufacturing the product.

- Use of non-contact sensors: In the traditional process the product or the part to be inspected is handled manually since it has to be positioned properly for inspection on the desk or suitable location. In CAQC non-contact sensors are used for the inspection purpose and they inspect the product without coming in contact with the product. The non-contact sensors operated by the computer are kept along the production line and they can check the product very quickly in the fraction of seconds. In future with further advancements in the technology, the robots would be used to carry-out the inspection process thus further automating and speeding the process.

Computer-aided Maintenance and Reliability Management Systems for Conveyor Belts

Modern transport systems based on belt conveyors are often very complex, in terms of both technology and design, which translates directly into their costliness. At the same time, the performance of a conveyor-based system determines not only the efficiency of a company's entire transportation system, but also the efficient and effective operation of other, cooperating subunits of a company's production system. In many cases, belt conveyor transport and cooperating subsystems have a serial structure. This means that a failure of a belt conveyor, which is one of the basic components of a transport system of this type, brings all other machinery and equipment to a stall. Therefore, the problem of ensuring an adequate level of availability of a continuous transport system is extremely important. Availability, in the case of belt conveyors, depends not only on their usability, as determined by their design and manufacture, but also on the proper level of utilization, understood, among others, as the degree of complexity and effectiveness of the methods and tools employed for their industrial diagnosis.

Diagnostic Systems in the Maintenance of Conveyor Systems for Internal Transport

The application of methods of technical diagnosis in the maintenance of belt conveyors is becoming an important issue from the point of view of effective utilization of those machines. However, methods of technical diagnosis are still underused in the case of belt conveyors and internal transport systems based on them. This is particularly important as modern means of information transfer make it possible to effectively monitor the state of machinery and equipment, and other computer-aided methods and techniques enable processing of the data collected using inference and prediction methods.

Typical monitoring of the condition of a machine involves measurement of the physical parameters of the work process that guarantee correct and safe operation and user safety. In the case of belt conveyors, monitoring typically encompasses the following parameters or characteristics: belt speed, belt run-off, belt wear, temperature of the elements that cooperate with the belt, the filling status of the transfer section, and detection of metal components in the material conveyed, rotational speed of the drum, and conveyor performance. In general terms, then, monitoring of belt conveyors relates to diagnostic symptoms and technological, operating and safety parameters. The current condition of a belt conveyor, in the context of effective management of this means of transport, and especially its belting, can be determined using a variety of methods, tools and systems. The most effective and efficient among those is diagnosis of the elements of a belt transport system. Therefore, in modern conveyor systems, the parameters that characterize the operation of the whole system, the conveyor, or the belt itself are usually monitored continuously using mobile or stationary measuring systems. In the case of mobile systems, the results of measurements can indicate that the conveyor operates properly, but they may also point to errors in design assumptions, calculations and selection of conveyor components, which result in decreased durability of the conveyor and its components as well as a lower reliability of the whole structure. Another group of diagnostic methods make use of computer techniques to control a single parameter or component whose condition has a significant impact on the reliability of the whole device and the entire belt conveyor system. A good example is a method used for monitoring the core of a steel cord conveyor belt that utilizes an apparatus for magnetic monitoring of damage. This system is used to monitor the condition of the belt core directly on the conveyor belt, and it works by measuring and recording changes in the magnetic field resulting from discontinuities in the belting, i.e. locations at which the steel cords of the core have been damaged. Another system that works in a similar manner is a system for monitoring and detection of damage of reinforced conveyor belts. The system generates a static magnetic field with a circuit that is closed by the reinforcing cords at the measuring sections along the belt and in the plane perpendicular to its surface. The system works by measuring magnetic field strength, processing the measurement signal, and analyzing the damage by means of an electronic data processing technique. Other diagnostic methods include radiographic examinations of fabric and steel cord belting for breaks and corrosion and ultrasonic rip detection.

One of the few methods that have so far been developed to prevent the occurrence of some of the types of belt failures is one in which special sensors are installed in the belting which signal belt damage or wear; these sensors work on the premise of transponders – electronic devices used in active radar systems. A transponder receives radio signals from one (transmitting) system and then collects processes and amplifies them to transmit to another (receiving) system. A system that works in a similar way is Sensor Guard. However, it only serves to protect belting against extensive rip damage and signals lateral movement and excessive slippage of the belt. In modern transport systems, it is a common trend to install complete computer-assisted monitoring systems.

One such system is ZEFIR NT that is used to monitor a system for transporting rock materials. Its task is to collect, analyze and visualize data related to the operation time of the individual devices, the number of start-ups, the length of outage periods, etc. Unfortunately, this system does not perform any diagnostic functions, and the data it generates can only be analyzed with a considerable delay.

The systems are used to control certain parameters related to the failure rate of conveyor belts, but they do not protect belts against sudden breaking in the area of the splice. They do not represent comprehensive monitoring and control systems, and the data collected with their help are often analyzed manually. For these reasons, it is necessary to develop appropriate, fully automated systems for controlling internal transport, including conveyor-based transport. A number of types of damage that may be incurred by a conveyor belt and its joints necessitate the use of advanced and maximally versatile diagnostic systems. Efficient and effective diagnosis of belt conveyors is also essential for continuous assessment of the state of belting aimed at preventing major failures and thus prolonging the conveyor's service life.

This means that it is necessary to implement into industrial practice a fully automated system for diagnosing internal conveyor transport, especially that the costs incurred as a result of losses due to belt damage are still very high. The use of simplified, manual analyses is time consuming and inefficient, especially in the era of the dynamic development of computer maintenance management systems (CMMs). Inefficient diagnosis may lead to reduced availability of a transport system; extend repair and maintenance time and increase maintenance costs. This is especially important when a conveyor belt breaks at the splice. Research on the design of belt conveyors and the structure of their load-bearing elements pays insufficient attention to conveyor belt splices, whose strength significantly affects the durability and reliability of conveyors, and even the entire systems that they are part of. A modem, RAM, and a microprocessor constitute its control system. The data transmission and processing system comprises a personal computer and phone lines or network cabling. Measurement points are marked with permanent magnets embedded in the conveyor belt on opposite sides of each splice. The location of the magnets during conveyor movement is identified by a semiconductor magnetic field sensor.

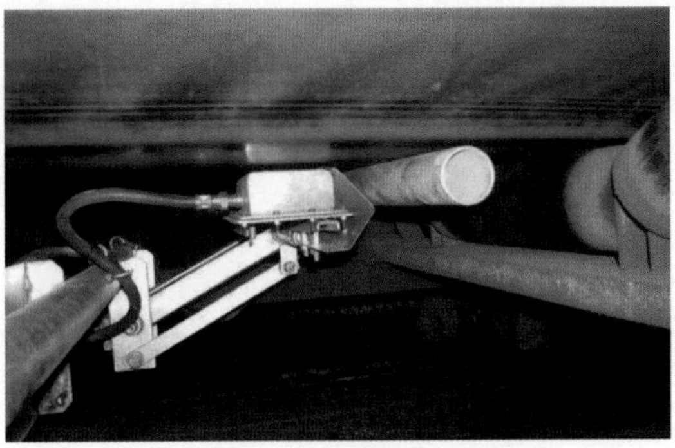

Measuring–diagnostic system.

The principle of operation of the system is based on the appropriate use of the magnetic sensor located under the moving conveyor belt. Magnetic field strength increases as the magnet approaches the sensor; and when the magnet passes the sensor, the sense of magnetic field strength changes. At this point, the strength of the magnetic field changes the most in the shortest time. A pulse registration method that works in this way makes it possible to measure the time between two measuring points at a high accuracy. The measurement data obtained are sent to a computer equipped with suitable software for acquisition and analysis of data and visualization of results. The original monitoring system designed on this premise can signal in advance the occurrence of conditions that normally accompany an impending break in a belt splice. The monitoring system is designed in a way that endows it with powerful data analysis capabilities and offers additional benefits, allowing a specialist to:

- Analyze those maintenance parameters of the belt conveyor that contribute to frequent failures due to belt breaks and to develop an appropriate intelligent system for monitoring the operating conditions of the device and for its automatic control aimed at eliminating critical situations;
- Analyze, on the basis of the results of measurements conducted under actual conditions of use, the durability of different types of bonded joints used in conveyor belts to enhance the quality of the joints and develop or modify the technology of their manufacture;
- Carry out analyses of the causes of defects in the manufacture of belt splices;
- Make a practical comparison of the peel strength and shear strength properties of adhesive materials used for splicing belts.

Measurement of the length of a conveyor belt splice begins when the magnetic field sensor sends a message that it has detected a reference point. This point marks the beginning of the measurement section and it resets the pulse counter. Splice length is measured indirectly. The moving belt turns a roller which is located under the upper section of the conveyor belt in such a way that there is permanent pressure between the roller and the moving belt, which prevents slippage between these two elements. Attached to the roller is a disk with evenly spaced indentations around its circumference. The pulse counter system uses a slotted optocoupler, which is a system that consists of two photoelements. One of them emits light beams (this is usually an LED), and the other is a photodetector, which converts the incoming light beams into electric voltage. The photodetector is a photodiode or a phototransistor. One of the disk slots is positioned between the two elements of the optocoupler. If the element that finds itself in the slot of the disk is impervious to light beams, a change in the voltage at the photodetector occurs (the voltage may rise or fall, depending on the electric system in which the photodetector operates). The moving belt turns the roller, which has a defined, constant diameter. The disk mounted on the roller has 200 holes (slots) around its circumference. This allows one to determine the distance travelled by the conveyor belt over the time when the disk has rotated by an angle equal to the rotation of the disk by one slot.

The optocoupler is used to count the passing holes (slots). The impulses start to be counted when a signal is received from the magnetic field sensor that the reference point has been detected. The counter starts counting the impulses coming from the optocoupler. Measurement ends when the magnetic field sensor sends a signal that it has detected a second reference point, which marks the end of the measuring section, i.e. the joint or section of the belt measured at a given time. The value of the counter is stored in the RAM of the control system and the counter is reset to zero. A microprocessor located in the pulse counter circuit controls the flow of data between the optocoupler, and the control system.

An important element of the computer-assisted measurement system is its control system. The control system of the measurement system for monitoring instantaneous changes in the length of conveyor belt splices consists of three components – a microprocessor, RAM and a modem. The microprocessor controls data flow between the measurement system and the control system. Data incoming from the pulse counter circuit are transferred to RAM. Each measurement is stored according to a specified pattern. Readout of the pulse counter is saved along with the hour, minute, second and the date of receipt of the data by the control system. The control system has been equipped with a large storage capacity because of the substantial amount of incoming data and the high rate of transfer of these data by the user to a PC. The large number of measurement data recorded is a consequence of the high-speed at which the conveyor belt moves and the large number of splices, whose passage through the measuring system is recorded many times a day.

For measurements of this type, a separate analysis of static and dynamic errors has to be conducted. Static error is independent of the time-varying nature of the value measured and it is not associated with transient processes in the device. Since ideal static conditions do not exist in practice, it is assumed that changes in the value of the parameter measured are at the level of resolution of the measuring instrument. Static error of digital measurement is comprised of quantization error and analog error. Quantization error is a consequence of mapping an infinite set of continuous (analog) values onto a finite set of quantized digital values. Analog error is more complex. It consists of the threshold error of the system which is unable to respond to infinitely small changes in the parameter measured. Reference source error is usually the main component of analog error and it results from the imperfect nature of the reference. The last component of analog error is the error associated with noise and other types of interference in the measurement circuit.

Dynamic error arises when the value measured changes over time. Dynamic errors are the difference between the value measured and the actual value. The main components of this error are quantization errors associated with sampling and transient states of the measuring instrument. Dynamic error is produced when a continuous value is averaged to discrete time intervals, when the sampling frequency is too low, i.e. lower than the frequency of changes of the analog signal, and when the instrument is in a transient state, for example, as a consequence of a change in the direction of the magnetic field in the magnetic field sensor.

Digital filters used in the proposed measurement system are designed to smooth out the signal, separate the relevant components of the frequency spectrum, and determine parameters in different frequency ranges. When designing a digital filter, care should be taken that the sampling frequency of the filter is higher than any component frequency of the signal at the input of the ADC. Another assumption is that the sampling period must be shorter than any time constant in the filter. This is to prevent a situation in which the analog signal ceases to be approximated by the digital signal. A digital filter must be sufficiently precise, i.e. it must have a low degree of quantization. If the degree of quantization of a filter is greater than the input of the ADC upstream of the filter, some information will be lost due to slower data-processing upstream of the filter. An unwanted effect in digital filters is clipping of the signal. It can lead to oscillation in the filter, which is a cause of errors.

Another problem related to measurement are the distortions arising from the nature of the environment in which the measurement system works. They give rise to signals in the measurement circuit that are not the result of the measurement process. The distortions are formed as a result of electromagnetic fields acting on the measuring system. Systems that are most liable to disturbances are those located in the vicinity of large electrical machines and those that are surrounded by a dense network of electric wires, as is usually the case with measurements made in industrial environments. Particularly sensitive to disturbance are analog systems. The continuous output signal can easily be deformed, especially when the voltage in the analog circuit is low, e.g., several millivolts. In this case, a distortion of the order of one millivolt can lead to an error between 10% and 20%. Measurement systems can be protected against this type of interference by shielding.

Computer-assisted Analysis of Measurement Data and Computer-aided Maintenance

As noted, among others, by Kaźmierczak, the decision-making process for maintenance activities can and should be effectively aided by using appropriate computer tools. One type of such tools is Computer-aided Maintenance Management Systems (CMMS), which are developed to perform functions that go beyond typical database maintenance. An optimal CMM system should also encompass elements of the decision making process, including, among others.

- Effective collecting, processing and sharing of data stored in database systems using diagnostic methods for evaluating the state of an object,

- Development of a knowledge base to complement the diagnostic data,
- Designing an advisory system tool to support decisions on the state of the object using the information stored.

Identical assumptions were adopted for the design of software modules for the measuring system. The main functions of this software are recording and processing of process data, control of measuring equipment and communication with the user. This is why in this program the tool changing column has been placed beside the data window. The tool buttons are used to activate various functions of the program, such as setting the compensation of the magnetic sensor or importing data to the user's computer. The program that supports the measurement system offers, among others, the possibility of visualizing changes in the length of a particular splice during operation of a conveyor, visualizing changes in the length of each section of the belt between the joints and collective presentation of these values for all (or selected) splices of the conveyor and all sections of the belt between the splices. It also enables statistical analysis of the values recorded and export of measurement data, graphics, etc.

The measurement data and their analysis using the proposed computer program allow visualizing and comprehensively analyzing measurement results, and the data obtained in this way allow to continuously monitoring the status of all splices and sections of the belt between them. The program can signal an alarm condition and helps evaluate the operating conditions of the conveyor, identify a single splice any time during the operation of the conveyor belt, and assess the effectiveness of maintenance work – for example, strengthening of a joint with mechanical components. This means that the measurement system can signal in advance the occurrence of conditions that normally accompany an impending break in a belt splice. The appropriate design of the system and the dedicated program for data analysis also afford additional benefits arising from the application of the system, which include, among others:

- Continuous measurement of changes in the length of all splices and belt sections between them to within 1 mm;
- About 1500 measurements per day for each conveyor, depending on the number of its splices;
- Real-time visualization of changes in splice length;
- Identification of splices and pinpointing of their exact location on the conveyor any time during its operation;
- Analysis of historical data – the possibility of collecting information on each splices and belt section (date and place of manufacture, manufacturer's identity, operation time, load, place of operation, etc.);

- The possibility of integrating the system as a module into other diagnostic–visualization systems, especially in self-management systems;
- Availability of accurate information on the operating time of the conveyor, its instantaneous speed values, downtime periods, number of stops and times of restart, and the precise length of the conveyor belt at any time of its operation;
- Data processing enabling predictive control of the operation of the conveyor belt.

It is due to these features that the system can be classified as a diagnostic system of the second level of security of the transport process, which is a higher level. Not without significance is also the possibility the system gives of long-term analysis of historical data for every single joint that forms part of the conveyor belt from its manufacture until the end of its useful life.

According to the concept of CMM, the monitoring device can also be transformed into an intelligent machine that has the ability to independently respond to the changing operating conditions and to eliminate conditions that cause belt breaks by anticipating future operating parameters and the consequences of their occurrence.

As it follows from previous analyses, both quality- and quantity oriented knowledge of changes (a decrease) in the value of strength parameters of belt sections and splices during the entire maintenance cycle is indispensable when it comes to choosing a belt for a given conveyor and predicting the optimum time point for performing maintenance, based on the precisely specified operation time and load distributions. It may also be useful for the assessment and verification of newly developed mathematical models of operation time distribution.

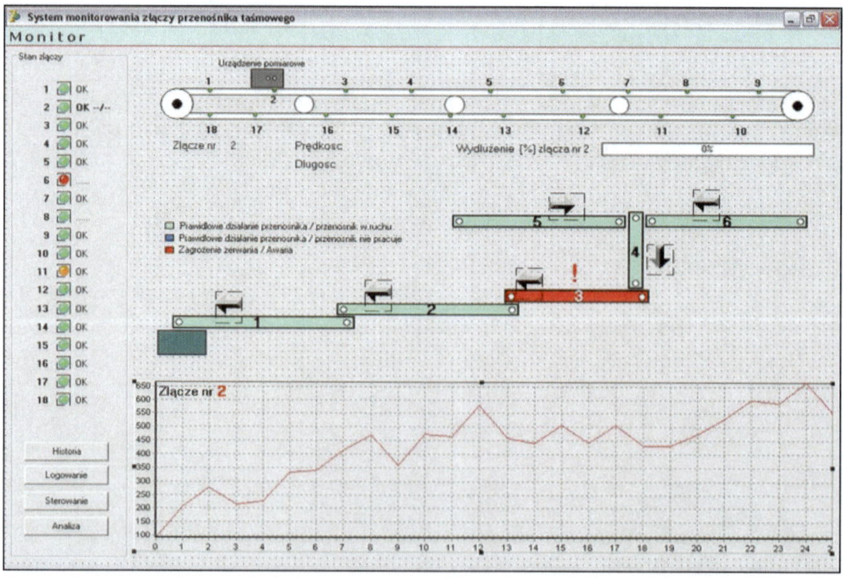

A conception of a method for visualizing data from a system for monitoring conveyor belt joints.

Therefore, bearing in mind that conveyor belts are commonly used in internal transport systems, and that there are no effective monitoring and control systems for preventing failures of these devices caused by splice breaks, it can be concluded that the results of the present study will not only increase the knowledge and experience of industrial use of diagnostic and monitoring systems, but will also reduce losses associated with downtime due to unexpected failures.

It can thus be stated that the objective of the present study is to design not only a comprehensive dispatching monitoring system, but also a smart actuator, which, when integrated into currently used transport monitoring and control systems, will allow to effectively expand them and make fuller use of diagnostics in self-management systems. The measurable benefits of implementing this concept of monitoring and controlling of belt conveyors include, among others:

- The possibility of constant surveillance of all conveyor belt splices during operation, using a non-invasive method;
- Identification of joints and assessment of their durability and reliability throughout their operating life, regardless of their current location;
- Real-time data mining of monitoring data along with a comprehensive analysis of historical data, which are especially useful in belt management;
- Simple design and easy operation that does not require outsourcing;
- The possibility of assessing the effectiveness of maintenance activities.

The conception of a system for monitoring conveyor belt joints developed in this study corresponds, in accordance with the theory of diagnosis, to real time diagnosis of an object with simultaneous signalling and visualization of its state. Implemented in this way, the monitoring method yields a real-time diagnostic system for conveyor transport. This system collects during its operation a considerable amount of different types of data, which provide information on the object being diagnosed. This excess of information, which operators have to analyze using visualization of the current parameters characterizing the analyzed object, may eventually lead to an information overload, often resulting in ignoring the relevant information or a failure to use the potential that measurement data collected by the system over a long time have. The monitoring system itself, then, despite a number of undeniable advantages, is not, in the long run, a sufficient solution to the optimal use of information it generates.

That is why it has to be expanded to become a fully automated computerized system for supporting the diagnosis of belt conveyors. Its primary role should be to process collected data and analyze them to support the decisions made by supervisory personnel or to take such decisions automatically, as far as possible, and to utilize them for current or temporary control of the transport system.

In this way, the system for monitoring conveyor belt splices and the belting between them would become a system of continuous surveillance (supervision and safeguarding) of the state of the analyzed object. Depending on its application, it could be used to support decision-making to ensure proper operation of the object or to control the object to prevent its failure and the consequences thereof.

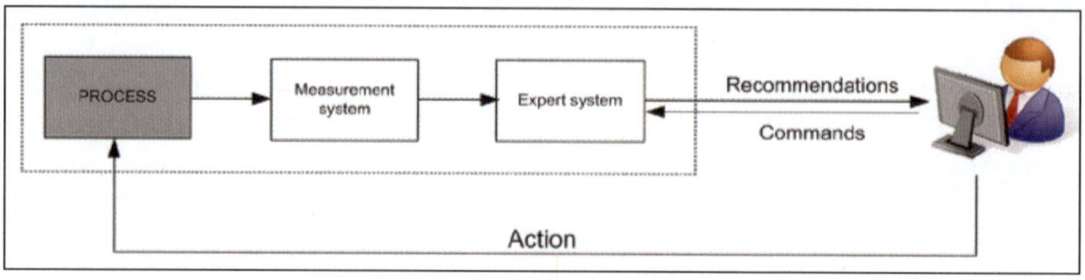

The concept of an expert decision support system for conveyor belt transport.

This is in line with diagnostic terminology, in which supervision stands for monitoring of an object and taking steps to maintain its proper operation, and safeguarding means the actions and technical measures taken to eliminate a potentially dangerous course of a process or to prevent the effects of such a course.

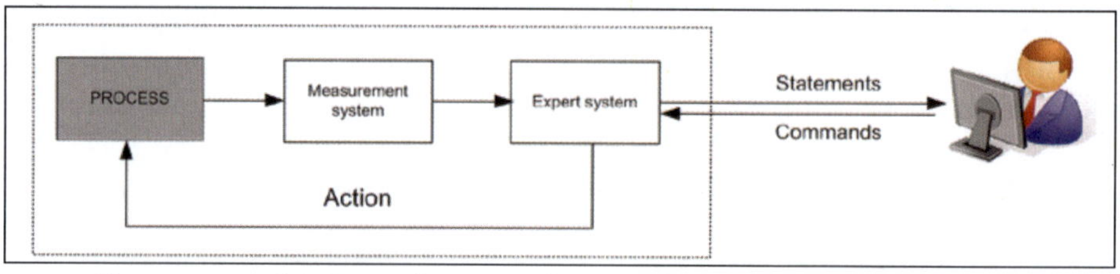

The concept of a preventive control system for conveyor belt transport.

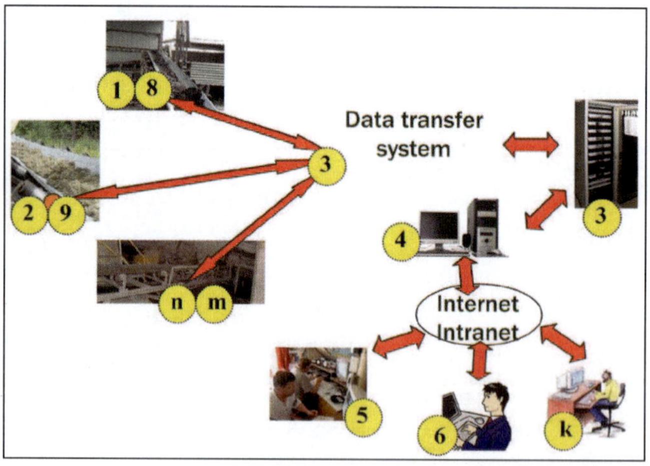

A schematic diagram of the concept of a comprehensive system for monitoring belt conveyor transport: 1, 2,..., n – measuring devices; 3 – a long-distance data transmission system; 4 – software for real time analysis and collecting of measurement data, processing of historical data and controlling the system; 5, 6,..., k – end-user interfaces compatible with the company's dispatching system; 8, 9,..., m – signalling devices.

Therefore, based on the results of the work carried out so far, it is also possible to build a comprehensive system of continuous surveillance of conveyor belts that could be used in company transport systems. Such a system would combine diagnostic equipment for analyzing the reliability-relevant operation parameters of the individual conveyors with elements of a computer-based advisory and control system.

The primary objective of the system will be to automatically monitor the status of all bonded joints to prevent their rupture, and to track the operation history of all belts and their individual sections. A system of this kind is versatile in that it can be used in different types of companies, in accordance with their specific branch-related characteristics and requirements. Thanks to its versatility and the flexibility of its software modules, the system can be combined with other dispatching diagnostic and monitoring systems.

One problem related to operating this type of system is that it requires collection and processing of various types of data, which are often difficult to describe or analyze in real-time. It would therefore be important to expand the proposed splice monitoring system to a diagnostic advisory system forming part of a comprehensive diagnostic system based on intelligent tools.

Computer-aided Manufacturing

Computer-aided manufacturing (CAM) is an application technology that uses computer software and machinery to facilitate and automate manufacturing processes. CAM is the successor of computer-aided engineering (CAE) and is often used in tandem with computer-aided design (CAD).

In addition to materials requirements, modern CAM systems include real-time controls and robotics.

CAM reduces waste and energy for enhanced manufacturing and production efficiency via increased production speeds, raw material consistency and more precise tooling accuracy.

CAM uses computer-driven manufacturing processes for additional automation of management, material tracking, planning and transportation. CAM also implements advanced productivity tools like simulation and optimization to leverage professional skills.

Depending on enterprise solution and manufacturer, CAM may present inadequacies in the following areas:

- Manufacturing process and usage complexity.
- Product Lifecycle Management (PLM) and modern enterprise integration.
- Machine process automation.

Modern CAM solutions are scalable and range from discrete systems to multi-CAD 3D integration.

Computer aided manufacturing typically uses software to translate drawings and data into detailed instructions that can drive some sort of automated tool. As an example, a 2D digital drawing can be used to guide a laser or physical cutting tool to cut cladding or other components to fit an architect's design.

As the Siemens definition points out, the programming language generated from the drawing or other data set that is then used to control the machine tool is referred to within the industry as the G Code. This G-code tells the tool how to make something by telling the motors where to move, how fast to move, and what path to follow.

The Benefits of CAM

Using CAM has a number of benefits when it comes to creating components used in building construction. Compared to manually operated machines, CAM generally offers:

- Greater speed in producing components.
- Greater accuracy and consistency, with each component or finished product exactly the same.
- Greater efficiency as computer controlled machines does not need to take breaks.
- High sophistication in terms of following complex patterns like tracks on circuit boards.

There are some limitations. CAM-enabled machines are generally designed for a particular task and are not incredibly versatile, although new systems and designs are emerging all the time.

They also need an upfront investment and skilled operators and programmers. Once in place, however, they could potentially bring large savings in time and efficiency, thereby reducing costs and saving companies thousands.

Computer-aided Design

Computer-aided design (CAD) involves creating computer models defined by geometrical parameters. These models typically appear on a computer monitor as a three-dimensional representation of a part or a system of parts, which can be readily altered by changing relevant parameters. CAD systems enable designers to view objects under a wide variety of representations and to test these objects by simulating real-world conditions.

Computer-aided manufacturing (CAM) uses geometrical design data to control automated machinery. CAM systems are associated with computer numerical control (CNC) or direct numerical control (DNC) systems. These systems differ from older forms of numerical control (NC) in that geometrical data are encoded mechanically. Since both CAD and CAM use computer-based methods for encoding geometrical data, it is possible for the processes of design and manufacture to be highly integrated. Computer-aided design and manufacturing systems are commonly referred to as CAD/CAM.

CAD had its origins in three separate sources, which also serve to highlight the basic operations that CAD systems provide. The first source of CAD resulted from attempts to automate the drafting process. These developments were pioneered by the General Motors Research Laboratories in the early 1960s. One of the important time-saving advantages of computer modeling over traditional drafting methods is that the former can be quickly corrected or manipulated by changing a model's parameters. The second source of CAD was in the testing of designs by simulation. The use of computer modeling to test products was pioneered by high-tech industries like aerospace and semiconductors. The third source of CAD development resulted from efforts to facilitate the flow from the design process to the manufacturing process using numerical control (NC) technologies, which enjoyed widespread use in many applications by the mid-1960s. It was this source that resulted in the linkage between CAD and CAM. One of the most important trends in CAD/CAM technologies is the ever-tighter integration between the design and manufacturing stages of CAD/CAM-based production processes.

The development of CAD and CAM and particularly the linkage between the two overcame traditional NC shortcomings in expense, ease of use, and speed by enabling the design and manufacture of a part to be undertaken using the same system of encoding geometrical data. This innovation greatly shortened the period between design and manufacture and greatly expanded the scope of production processes for which automated machinery could be economically used. Just as important, CAD/CAM gave the designer much more direct control over the production process, creating the possibility of completely integrated design and manufacturing processes.

The rapid growth in the use of CAD/CAM technologies after the early 1970s was made possible by the development of mass-produced silicon chips and the microprocessor, resulting in more readily affordable computers. As the price of computers continued to decline and their processing power improved, the use of CAD/CAM broadened from large firms using large-scale mass production techniques to firms of all sizes. The scope of operations to which CAD/CAM was applied broadened as well. In addition to parts-shaping by traditional machine tool processes such as stamping, drilling, milling, and grinding, CAD/CAM has come to be used by firms involved in producing consumer electronics, electronic components, molded plastics, and a host of other products.

Computers are also used to control a number of manufacturing processes (such as chemical processing) that are not strictly defined as CAM because the control data are not based on geometrical parameters.

Using CAD, it is possible to simulate in three dimensions the movement of a part through a production process. This process can simulate feed rates, angles and speeds of machine tools, the position of part-holding clamps, as well as range and other constraints limiting the operations of a machine. The continuing development of the simulation of various manufacturing processes is one of the key means by which CAD and CAM systems are becoming increasingly integrated. CAD/CAM systems also facilitate communication among those involved in design, manufacturing, and other processes. This is of particular importance when one firm contracts another to either design or produce a component.

Use of CAD

CAD is used to accomplish preliminary design and layouts, design details and calculations, creating 3-D models, creating and releasing drawings, as well as interfacing with analysis, marketing, manufacturing, and end-user personnel.

CAD facilitates the manufacturing process by transferring detailed information about a product in an automated form that can be universally interpreted by trained personnel. It can be used to produce either two-dimensional or three-dimensional diagrams. The use of CAD software tools allow the object to be viewed from any angle, even from the inside looking out. One of the main advantages of a CAD drawing is that the editing is a fast process as compared to manual method. Apart from detailed engineering of 2D or 3D models, CAD is widely used from conceptual design and layout of products to definition of manufacturing of components. CAD reduces design time by allowing precise simulation rather than builds and test physical prototypes. Integrating CAD with CAM (Computer Aided Manufacturing) streamlines the product development even more.

CAD is currently widely used for industrial products, animated movies and other applications. A special printer or plotter is usually required for printing professional design renderings. CAD programs use either vector-based graphics or raster graphics that show how an object will look.

Types of Computer-aided Design

There are many kinds of CAD, covering a vast range of uses. However, in each case, the design of their virtual components requires a different approach.

At the lower end of the 2D systems, there are several free and open source systems. In other words, there are many freely-available programs which we may modify and redistribute.

These programs provide an approach to drawing without all the intricacies regarding scale and placement on the drawing sheet that hand drafting requires. During the creation of the final draft, the designer can make adjustments.

3D Wireframe

3D Wireframe is an extension of 2D drafting. The designer must manually insert each line into the drawing. The finished product has no mass properties associated with it. For example, it is not possible to add features, such as holes, directly to it.

The 3D Wireframe model is an edge or skeletal representation of a real-life object. Models consist of lines, points, arcs, circles, and other curves that define the center lines or edges of objects.

3D 'Dumb' Solids

With 3D 'Dumb' Solids, the designer creates things in a way similar to manipulations of real world objects. You can add cylinders, prisms, spheres, and other basic 3D geometric forms to the design. You can also take them away as if you were cutting or assembling real-world objects.

Designers can also generate 2D projected views from the models.

More upmarket systems offer capabilities to incorporate more ergonomic, aesthetic, and organic features into designs.

With a computer-aided design program, you can visualize where to place goods on the shelves of supermarkets and other stores. You can move them around and even look at them from different angles.

Advantages and Disadvantages

Modeling with CAD systems offers a number of advantages over traditional drafting methods that use rulers, squares, and compasses. For example, designs can be altered without erasing and redrawing. CAD systems also offer “zoom” features analogous to a camera lens, whereby a designer can magnify certain elements of a model to facilitate inspection. Computer models are typically three dimensional and can be rotated on any axis, much as one could rotate an actual three dimensional model in one’s hand, enabling the designer to gain a fuller sense of the object. CAD systems also lend themselves to modeling cutaway drawings, in which the internal shape of a part is revealed, and to illustrating the spatial relationships among a system of parts.

To understand CAD it is also useful to understand what CAD cannot do. CAD systems have no means of comprehending real-world concepts, such as the nature of the object being designed or the function that objects will serve. CAD systems function by their capacity to codify geometrical concepts. Thus the design process using CAD involves transferring a designer’s idea into a formal geometrical model. Efforts to develop computer-based “artificial intelligence” (AI) have not yet succeeded in penetrating beyond the mechanical—represented by geometrical (rule-based) modeling.

Other limitations to CAD are being addressed by research and development in the field of expert systems. This field is derived from research done in AI. One example of an expert system involves incorporating information about the nature of materials—their weight, tensile strength, flexibility, and so on—into CAD software. By including this and other information, the CAD system could then “know” what an expert engineer knows when that engineer creates a design. The system could then mimic the engineer’s thought pattern and actually “create” more of the design. Expert systems might involve the implementation of more abstract principles, such as the nature of gravity and friction, or the function and relation of commonly used parts, such as levers or nuts and bolts. Expert systems might also come to change the way data are stored and retrieved in CAD/CAM systems, supplanting the hierarchical system with one that offers greater flexibility. Such futuristic concepts, however, are all highly dependent on our abilities to analyze human decision processes and to translate these into mechanical equivalents if possible.

One of the key areas of development in CAD technologies is the simulation of performance. Among the most common types of simulation are testing for response to stress and modeling the process by which a part might be manufactured or the dynamic relationships among a system of parts. In stress tests, model surfaces are shown by a grid or mesh, that distort as the part comes under simulated physical or thermal stress. Dynamics tests function as a complement or substitute for building working prototypes. The ease with which a part’s specifications can be changed facilitates the development of optimal dynamic efficiencies, both as regards the functioning of a system of parts and the manufacture of any given part. Simulation is also used in electronic design automation, in which simulated flow of current through a circuit enables the rapid testing of various component configurations.

The processes of design and manufacture are, in some sense, conceptually separable. Yet the design process must be undertaken with an understanding of the nature of the production process. It is necessary, for example, for a designer to know the properties of the materials with which the part might be built, the various techniques by which the part might be shaped, and the scale of production that is economically viable. The conceptual overlap between design and manufacture is suggestive of the potential benefits of CAD and CAM and the reason they are generally considered together as a system.

Recent technical developments have fundamentally impacted the utility of CAD/CAM systems. For example, the ever-increasing processing power of personal computers has given them viability as a vehicle for CAD/CAM application. Another important trend is toward the establishment of a single CAD-CAM standard, so that different data packages can be exchanged without manufacturing and delivery delays, unnecessary design revisions, and other problems that continue to bedevil some CAD-CAM initiatives. Finally, CAD-CAM software continues to evolve in such realms as visual representation and integration of modeling and testing applications.

Computer-aided Technologies

Computer-aided technologies also known as Cax, applies different computer technologies to aid in the design, analysis, and manufacturing of products.

Integration of Computer-aided Technologies

The major objectives of computer-aided technology (CAx) are to simplify and to improve human's work (engineer, architect, physician, surgeon, etc.), by using the computer as an indispensable tool to solve a problem in a certain field (engineering and production, medicine, architecture, business, teaching, economy, etc).

The advanced computer-aided technologies (CAx) are focused on solving specific problems by increasing human's creativity and innovation obtained through collecting, using, and sharing information between interdisciplinary teams.

Computer-aided technologies in X field are general terms to define a technology, from a specific field of work, which is computed-aided. The substitute for X includes engineering (CAx-E), medicine (CAx-M), natural science (CAx-S), education (CAx-Ed), etc.

Nowadays, computer-aided technologies are not islands of automation, being integrated in general context of Lifecycle Management in X field. The concepts used to define the lifecycle management in engineering and medical field are the following:

- Product lifecycle management (PLM) in the industry field.

- Human lifecycle management (HUM) or health management across the human lifecycle.

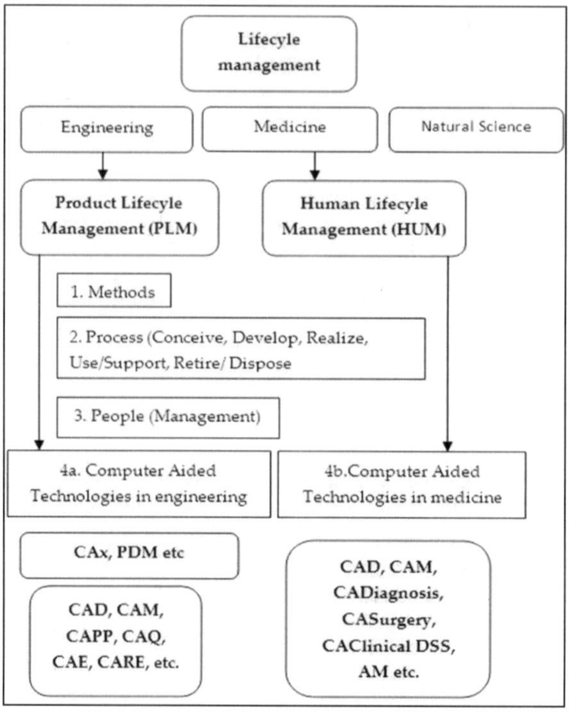

Schematic representation of lifecycle management in industry, medicine, and natural science.

Computer-aided Technologies in Product Lifecycle Management

Product lifecycle management (PLM) is the process of managing the entire lifecycle of a product including conception, design, manufacturing, quality control, use, service, and disposal of products, having integrated people, data, methods, CAx tools, processes, and business systems. PLM is a digital paradigm, products being managed with digital computer, digital information, and digital communication.

The main benefits of product lifecycle management (PLM) for the industry field are faster time-to-market, improved productivity, better product quality, decreased cost of new product introduction, improved design review, and approval processes, identifying potential sales opportunities and revenue contributions and reducing environmental impacts at end-of-life. PLM emerged from tools such as CAD, CAM, and PDM, being viewed as the integration of these tools with innovative technologies (e.g., additive manufacturing, reverse engineering), methods, people, and the processes through all stages of a product's life.

To improve the product lifecycle management, a lot of methods and techniques are used such as concurrent engineering, bottom-up design, top-down design, both-ends-against-the-middle design, design in context, design for X (DFX), TRIZ, lean production, design for six sigma (DFSS), total quality management (TQM), and failure mode effects analysis (FMEA).

Concurrent engineering or simultaneous engineering is a workflow that, instead of working sequentially through stages, carries out a number of tasks in parallel.

The most used design for X method is design for manufacture and assembly (DFMA) that is the combination of two methodologies: design for manufacture (DFM) and design for assembly (DFA), which mean the design of the parts for ease of manufacturing and the design of the product for ease of assembly.

Product data management (PDM) is the business function often within product lifecycle management (PLM) that is responsible for the management and publication of product data.

The tools used to access the information and knowledge regarding the product data is the computer-aided technologies (CAx). Computer-aided technologies (CAx) are the use of computer technology to aid in the design, analysis, production planning, manufacture of products, etc.

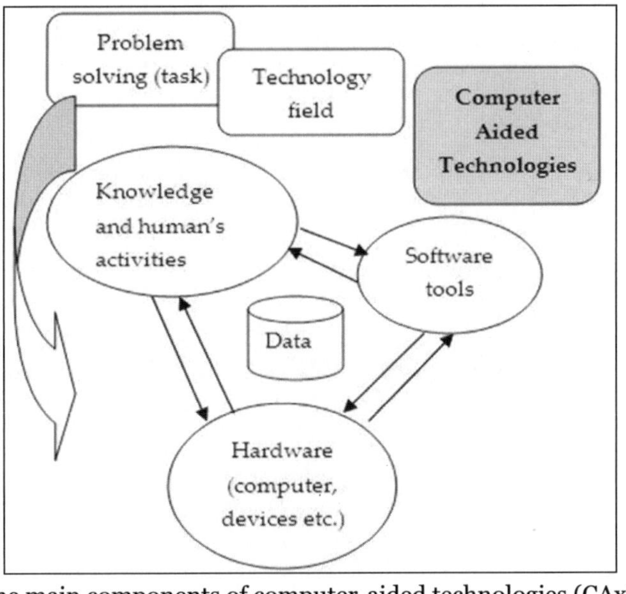

The main components of computer-aided technologies (CAx).

A CAx system can work like an "island of automation" or it can be integrated in the PLM system, interacting with other "islands of automation". Thus, the advanced CAx tools merge many different aspects of the product lifecycle management, including design, manufacturing, etc. CAx can be integrated, also, with other computational systems of management and planning of trials and output, such as MRP (material resource planning), ERP (enterprise resource planning), EDM (electronic document management), and PDM (product data management).

A CAx system may be defined, in generally, having the following main components:

- Hardware component consisting in computer and interactive devices,
- Software packages,
- Data,
- Knowledge and human's activities.

Computer-aided design (CAD), computer-aided engineering (CAE), computer-aided manufacturing (CAM), computer-aided process planning (CAPP), and computer-aided quality assurance (CAQA) are the most known and mature computer-aided technologies.

Computer-aided design is the computer-aided technology that involves the computer to assist in the creation, modification, analysis, and optimization of a design and design documentation.

Computer-aided engineering (CAE) is the computer-aided technology that involves the computer to analyze, simulate, and optimize the CAD geometry. CAE tools are available for a wide range of analyses: stress analysis, deformation, heat transfer, fluid flow, magnetic field distribution, kinematics, and dynamic analysis, etc.

Computer-aided manufacturing (CAM) is the computer-aided technology that involves the computer in planning, control, and management of manufacturing of any product. The most mature areas of CAM are the numerical control (NC) of the machine tools and programming of industrial robots that perform tasks as assembly, welding, etc.

The following are the most known commercial software tools for computer-aided technologies:

- CATIA by Dassault Systemes, Creo by PTC, NX by Siemens, PowerShape/PowerMill by Delcam etc., in the field of CAD/CAM.
- Materialise Magics and Netfabb Studio, in the field of 3D-printing/ additive manufacturing.
- RapidForm and Geomagic in the field of computer-aided reverse engineering.
- Ansys, Abaqus, COMSOL Multiphysics, Adams, LMS Virtual.Lab are focused on CAE.

Computer-aided Technologies in Health Management Across the Human Life Cycle

Medical technology is the type of technology which is used to extend and improve human life. Medical technology is used to diagnose infections, treat diseases, and to make research on diseases affecting humans. Computer-aided technologies play an important role in health management across the human life cycle. The main applications of computer-aided technologies in the medical field are the following:

- Computer-aided design (CAD).
- Computer-aided detection and diagnosis.
- Computer-aided medical interventions or computer-aided surgery.
- Computer-aided clinical decision.
- Computer-aided simulation.

CAD systems play an important role in medical applications, allowing to simulate and analysis, prosthesis design, surgical implant design, blood flow analysis, preoperative planning for surgical operations, and computer-assisted surgery.

Virtual reality (VR) enables physicians and surgeons to interact, manipulate, and simulate the geometric 3D CAD models of anatomical structures directly in a virtual environment, with 3D displays and haptic devices.

Rapid prototyping (RP), 3D printing (3DP), and additive manufacturing (AM) technologies allow us to obtain a real copy of anatomical structures, before a medical implant is inserted or a surgical procedure is performed.

Computer-aided Process Planning

Computer-aided process planning is a link between design and manufacturing in a computer-integrated manufacturing (CIM) environment. Commonly used by manufacturing engineers, CAPP can be used to develop a product manufacturing plan based on projected variables such as cost, lead times, equipment availability, production volumes, potential material substitution routings and testing requirements.

The process-planning problem comes out as a rigid hierarchical structure of tasks where the lowest levels are well-defined tasks, such as determination of machine and cutting parameters, while the top levels control, coordinate and manage the entire system. Such a top-down manner of coordination makes it a tightly coupled distributed decision-making situation. The complexity of CIM systems with hierarchical structures grows rapidly with the size of the system, making them very expensive. Also, assimilating a large array of knowledge sources to plan the activities is a major hurdle.

The generative approach to CAPP starts with the product design specifications and can generate a detailed process plan complete with machine settings. CAPP systems use design algorithms, a file of machine characteristics, and decision logic to build the plans. Expert systems are based on decision rules and have been used in some generative CAPP systems.

CAPP has recently emerged as the most critical link to integrated CAD/CAM system into inter-organizational flow. Main focus is to optimize the system performance in a global context. The essentiality of computer can easily be understood by taking an example, e.g. if we change the design, we must be able to fall back on a module of CAPP to generate cost estimates for these design changes. Similarly for the case of the breakdown of machines on shop floor. In this case, alternative process plan must be in hand so that the most economical solution for the situation can be adopted. Figure is one such representation, where setting of multitude of interaction among various functions of an organization and dynamic changes that takes place in these sub functional areas have been shown. Hence, the use of computer in process planning is essential.

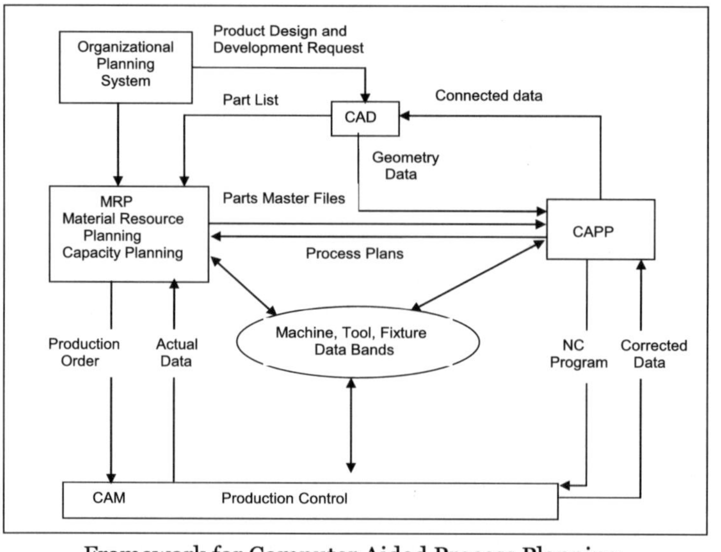

Framework for Computer Aided Process Planning.

CAPP is the application of computer to assist the human process planer in the process planning function. In its lowest form it will reduce the time and effort required to prepare process plans and provide more consistent process plan. In its most advanced state, it will provide the automated interface between CAD and CAM and in the process achieve the complete integration with in CAD/CAM.

Advantages over Manual Experience-based Process Planning

The uses of computers in process plan have following advantages over manual experience-based process planning:

- It can systematically produce accurate and consistent process plans.
- It leads to the reduction of cost and lead times of process plan.

- Skill requirement of process planer are reduced to develop feasible process plan.
- Interfacing of software for cost, manufacturing lead time estimation, and work standards can easily be done.
- Leads to the increased productivity of process planar.

With the emergence of CIM as predominate thrust area in discrete part industries process planning has received significant attention, because it is the link between CAD and CAM. Hence, computer aided process planning (CAPP) has become a necessary and vital objective of CIM system.

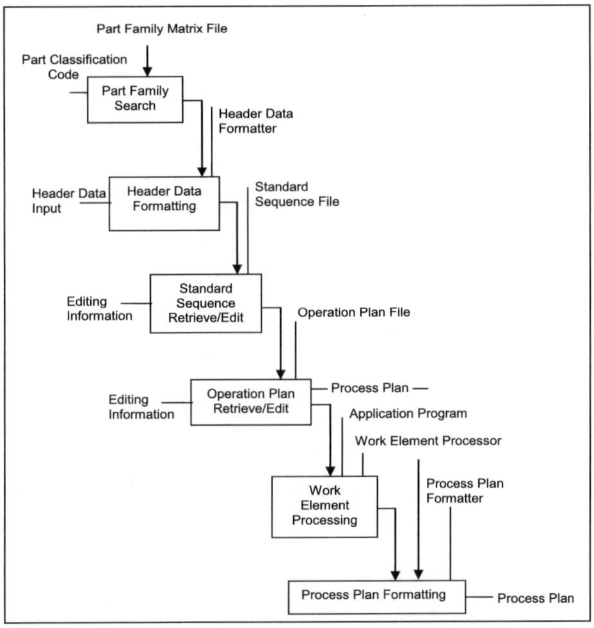

Flow Diagram of the CAPP Process Planning System.

Steps involved in CAPP

Now-a-days, rapid progress is being made in the automation of actual production process and also the product design element. However, the interface between design and production presents the greatest difficulty in accomplishing integration.

CAPP has the potential to achieve this integration. In general, a complete CAPP system has following steps:

- Design input,
- Material selection,
- Process selection,
- Process sequencing,

- Machine and tool selection,
- Intermediate surface determination,
- Fixture selection,
- Machining parameter selection,
- Cost/time estimation,
- Plan preparation,
- Mc tape image generation.

Approaches to Computer-aided Process Planning

In recent days, several computer-aided process planning systems are available for use for a variety of manufacturing operation.

These systems can broadly be clarified into two categories:

- Variant computer aided process planning method.
- Generative computer aided process planning method.

Variant Process Planning, Advantages and Disadvantages

Variant process planning approach is sometimes referred as a data retrieval method. In this approach, process plan for a new part is generated by recalling, identifying and retrieving an existing plan for a similar part and making necessary modifications for new part. As name suggests a set of standard plans is established and maintained for each part family in a preparatory stage. Such parts are called master part. The similarity in design attributes and manufacturing methods are exploited for the purpose of formation of part families. Using coding and classification schemes of group technology (GT), a number of methods such as coefficient based algorithm and mathematical programming models have been developed for part family formation and plan retrieval. After identifying a new part with a family, the task of developing process plan is simple. It involves retrieving and modifying the process plan of master part of the family.

The general steps for data retrieval modification are as follows:

Establishing the Coding Scheme

A variant system usually begins with building a classification and coding scheme. Because, classification and coding provide a relatively easy way to identify similarity among existing and new parts. Today, several classification and coding systems are commercially available. In some extreme cases, a new coding scheme may be developed.

If variant CAPP is preferred than it is useful for a company to look into several commercially available coding and classification systems (e.g. DCLASS, JD-CAPP etc.). Now, it is compared with companies before developing their own coding and classification system. Because using an existing system can save tremendous development time and manpower.

- Form the Part Families by Grouping Parts: The whole idea of GT lies into group numerous parts into a manageable number of part families. One of the key issues in forming part families is that all parts in the same family should have common and easily identifiable machined features. As a standard process plan are attached with each part family, thereby reducing the total number of standard process plans.
- Develop Standard Process Plans: After formation of part families, standard process plan is developed for each part families based on common part features. The standard plan should be as simple as possible but detailed enough to distinguish it from other.
- Retrieve and Modify the Standard Plans for New Parts: Step1 to step 3 are often referred as preparatory work. Each time when a new part enters the systems, it is designed and coded based on its feature, using the coding and classification scheme, and then assigned to a part family. The part should be similar to its fellow parts in the same family. Also, family's standard plan should represent the basic set of processes that the part has to go through. In order to generate detailed process routes and operation sheets to this part, the standard plan is retrieved from the data base and modified. Modification is done by human process planar. After this stage parts are ready for release to the shop.

The success of aforementioned process planning system is dependent on selection of coding scheme, the standard process plan and the modification process, because the system is generally application oriented. It may be possible that one coding scheme is preferable for one company and same is not for other company.

Due to use and advancement of computers, the information management capability of variant process planning is much superior. Otherwise it is quite similar to manual experience-based planning.

Advantages and Disadvantages of Variant CAPP

Following advantages are associated with variant process planning approach:

- Processing and evaluation of complicated activities and managerial issues are done in an efficient manner. Hence lead to the reduction of time and labour requirement.

- Structuring manufacturing knowledge of the process plans to company's needs through standardized procedures.
- Reduced development and hardware cost and shorter development time. This is an essential issue for small and medium scale companies, where product variety is not so high and process planner are interested in establishing their own process planning research activities.

Disadvantages of Variant Process Planning Approach

Following disadvantages are associated with variant process planning approach:

- It is difficult to maintain consistency during editing.
- Proper accommodation of various combinations of attributes such as material, geometry, size, precision, quality, alternate processing sequence and machine loading among many other factors are difficult.
- The quality of the final process plan largely depends on the knowledge and experience of process planner. The dependency on process planner is one of the major shortcomings of variant process planning.

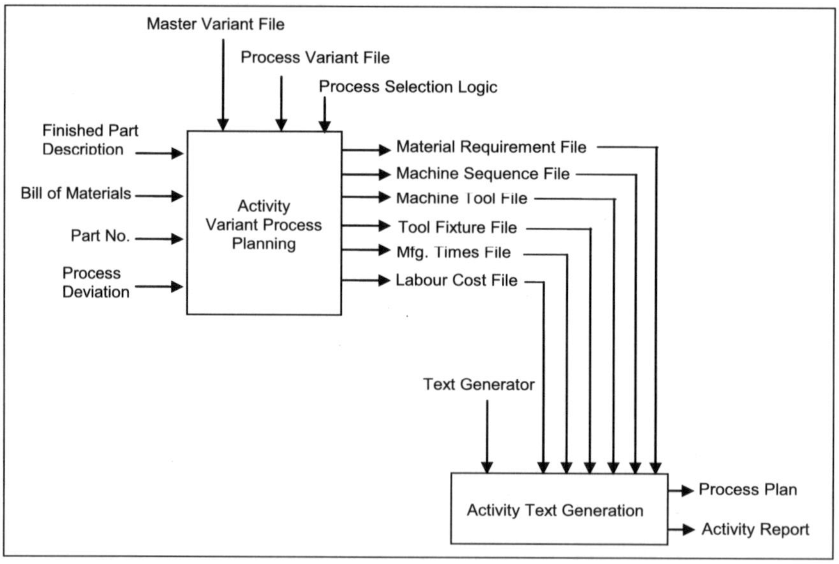

Framework of Variant Process Planning Activity.

Some of the most widely used process planning method developed by various companies are mentioned as follows:

- Mc Donnell-Douglas automation company under the direction and sponsorship of Computer Aided Manufacturing International (CAM-I) developed a system where CAPP can be used to generate process plan for rotational, prismatic and sheet metal part.

- Organization for Industrial Research (OIR) and General Electric Company have developed and another process plan named as MIPLAN. It accommodates both rotational and prismatic part, and is based on MICLASS coding.

Generative Process Planning, Advantages and Disadvantages

In generative process planning, process plans are generated by means of decision logic, formulas, technology algorithms, and geometry based data to perform uniquely processing decisions. Main aim is to convert a part form raw material to finished state. Hence, generative process plan may be defined as a system that synthesizes process information in order to create a process plan for a new component automatically.

Generative process plan mainly consists of two major components:

- Geometry based coding scheme.
- Proportional knowledge in the form of decision logic and data.

Geometry-based Coding Scheme

All the geometric features for all process such as related surfaces, feature dimension, locations, on the features are defined by geometry based coding scheme. The level of detail is much greater in generative system than a variant system.

For example, various details such as rough and finished state of the part are provided to transform into desired state.

Proportional Knowledge in the Form of Decision Logic and Data

Process knowledge in the form of decision logic and data are used for matching of part geometry requirement with the manufacturing capabilities.

Operation instruction sets are automatically generated to help the operators to run the machines in case of manual operation. NC codes are automatically generated, when numerically controlled machines are used.

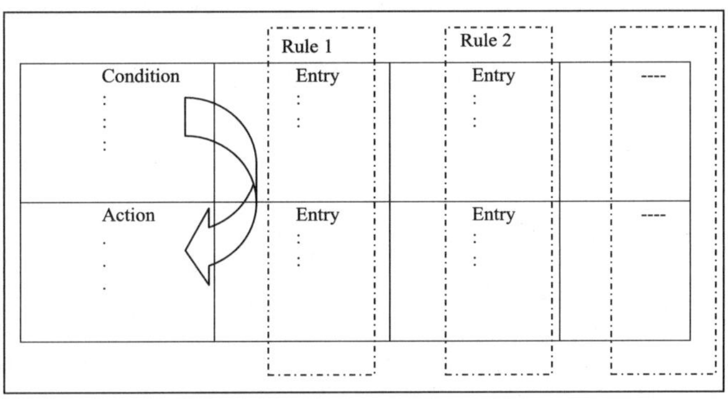

Framework of a decision table.

Manufacturing knowledge plays a vital role in process planning. The process of acquisition and documentation of manufacturing knowledge is a recurring dynamic phenomenon. In addition, there are various sources of manufacturing knowledge such experience of manufacturing personnel, handbooks, supplier of machine tools, tools, jigs and fixtures materials, inspection equipment and customers etc. Hence, in order to understand manufacturing information, ensuring its clarity and providing a framework for future modification, it is not only necessary but also inevitable to develop a good knowledge structure from wide spectrum of knowledge. Flowchart, decision trees, decision tables, algorithms, concepts of unit machined surfaces, pattern recognition techniques, and artificial intelligent based tools are used to serve the purpose.

The basic elements of decision tables are condition, action and rules. They are represented in the form of allocation matrix. Figure is one such representation where condition states the goal that we want to achieve and action states the operation that we have to perform. On the basis of experience the expert rules are formed by entry values to establish the relationship between condition and action.

Table: Boolean value-type entries.

Length of bar ≥ 8 in.	T*	F	
Diameter of bar 1 ≤ in.			
Diameter of bar ≥ 1 in	T		T
-	-	-	-
Extra Support	T		

Table: Continuous value-type entries.

Length of bar (in)		≤ 4	≥ 4	≤ 16	≥ 16
Diameter of bar (in.)	≤ 0.2	> 0.2	1> diameter >0.2	≥ 1	
Extra support	T		T		T

For a particular set of condition entries, look for its corresponding rule from that rule determine the action.

Advantages of Generative Process Plan

Generative process plans have a number of advantages. Among the major ones are the following:

- They rely less on group technology code numbers since the process, usually uses decision tree to categorize parts into families.
- Maintenance and updating of stored process plans are largely unnecessary. Since, any plan may be quickly regenerated by processing through the tree.

Indeed, many argue that with generable systems, process plans should not be stored since if the process is changed, and out-of-dated process plan might find its way back into the system.

- The process logic rules however must be maintained up to dated and ready for use. This provides the process planner with an assurance that the processes generated will reflect state-of-the-art technology.

Table: Description of various generative and variant and generative CAPP systems is mentioned table.

CAPP System	Part Shapes	Process Planning Approaches	Characteristics and Commercial Situation	Programming Languages Used
CMPP	Rotational	Generative	Uses English like language(COPPL)	FORTRAN 77
GENPLAN	All	Variant and Generative	Interfaced with CAD\ CAM	
GT-CAPP	All	Generative	Part family code used	
KAPPS	Rotational and Prismatic	Generative	Part family numbers used	LISP
MIPLAN	Rotational and Prismatic	Variant	Expert system based on MICLASS	
RTCAPP	Prismatic	Generative	Generic shell	
TURBOCAPP	Rotational	Generative	Knowledge based interfaced with CAD	PROLOG
XPLAN	All	Generative	Expert system based on DCLASS	FORTRAN 77
XPLAN-R	Rotational	Generative	Expert system based on DCLASS	FORTRAN 77
XPLANE	Rotational	Generative	Knowledge based	FORTRAN
XPS-1	All	Variant and Generative	COPPL used	FORTRAN

Knowledge-based Process Planning

The main forces behind to apply knowledge-based (KB) techniques for CAPP is the requirement of large amount of human expertise in CAPP. Based on the previous discussion, one realizes that a productive CAPP system must contain tremendous amount of knowledge – facts about the machine and shop environment as well as rules about sequencing machining operations must be included. A traditional CAPP program cannot learn new knowledge without a programmer explicitly rewriting it. The rigidity of traditional methodology endangers the implementation of CAPP systems. A KB system stores knowledge in a special manner so that it is possible to add, delete and modify facts and rules in the knowledge base without rewriting the program, i.e. it learns new things according to embedded learning procedures.

A complete set of manufacturing knowledge is not equipped by any existing knowledge-based process planning system. Most of these systems focus on a small portion of the issues in the domain of automated process planning using an expert systems approach. Some of them are:

EXCAP Family of Process Planning Systems

EXCAP, EXpert Computer-Aided Process Planning, developed by Davies and Darbyshire, is a knowledge-based system for rotational part process planning.

EXCAP-A and EXCAP-Y are previous generations of the current member of the EXCAP family of process planning systems.

GARI

GARI is the first AI-based CAPP program to appear in the literature. It is implemented in MACLISP and operates on CII-Honeywell Bull HB-68 computer under the MULTICS operating system. GARI utilizes production rules in its knowledge representation and generates a process plan from a model of the part. It emphasizes the "conflict resolution". The knowledge is rather subjective and specialized. As a result, in the planning process, "compromises are often necessary."

TOM: Techno Structure of Manufacturing

TOM is another production rule-based CAPP system written in PASCAL and runs on VAX computer. TOM was designed to accept input in two ways: (1) directly entering part desecration by the user, and (2) translating design data from COMPAC CAD system. TOM can deal with "holes" exclusively.

SIPP: Semi-Intelligent Process Planner

SIPP is an AI-based CAPP system for the creation of metal parts using chip metal removal operations. It is written in PROLOG and utilizes "frames" as its knowledge representation scheme instead of using production rules. Frames are used to represent two types of knowledge: (1) information about the characteristics of various kinds of machinable surfaces, and (2) the capabilities of various machining processes.

SIPS

SIPS, another AI-based CAPP system which selects machining operations for the creation of metal parts, is a successor to SIPP. It is written in LISP and is currently being integrated into the AMRF (Automated Manufacturing Research Facility) project, where it is used to select machining operations on a feature by feature basis.

Like SIPP, SIPS also employs branch and bound search strategy for the least-cost first solution in its inference engine. The basic difference between SIPP and SIPS is that SIPS used a new knowledge representation technique, called hierarchical knowledge clustering, instead of "flat" frames to represent problem-solving knowledge.

TOLTEC

TOLTEC is a system equipped with some learning capability. It takes input as feature-based part description interactively. The features are represented in a frame structure. It generates output in a form of operations and their sequence.

Turbo-CAPP

Turbo-CAPP is a knowledge-based CAPP system written in PROLOG and capable of:

- Extracting and interpreting surface features from a 2&1/2-D CAD data base.
- Performing intelligent reason for process planning.
- Learning new process and machining capabilities.
- Generating alternative process plans (based on the current status of the knowledge base).
- Creating generic NC part programs for automated.

Turbo-CAPP is designed to handle strictly symmetric rotational parts. It employs a backward chaining inference mechanism for plan generation. In the process of creating process plan and NC codes, the system must acquire knowledge from the user from time to time.

XPS-2 Family of CAPP Systems

CAM-I started the first structured development of process planning systems. It then embarked on a form-feature based, generative planning project, XPS and accomplished with the completion of XPS-2 in 1987. The form feature used to implement XPS-2 were taken from a "feature taxonomy" developed by CAM-I.

Other Knowledge-based CAPP Systems

Rather than aforementioned Knowledge-Based CAPP System some other KB process planning systems are in existence:

- CMPP is a planning system for planning cylindrical parts (also for some non-cylindrical features). It performs dimension, tolerance, and stock removal analysis based on a sophisticated algorithm with the objective of optimizing tolerance capabilities of shop equipment.

- Hi-MAPP developed by Brenfi and Khoshnevis.
- Wolfe and Kung in 1984 developed a CAPP system, which reads part geometry from a PADL model and generates process plans automatically.

Variant or Generative, Which to Use?

What CAPP approach (Variant or Generative) is better? This question has been constantly asked but, there is no definite answer to it.

Generally speaking, a variant system is better for manufacturing setting where similar parts are manufactured repetitively. Because parts are similar, Group Technology can easily be implemented and shows quick and significant return on investment (ROI). Because similar parts are produced repetitively, process plan can be retrieved, slightly modified and used, without going through too much trouble. On the other hand, generative process planning is better suited for a manufacturing environment in which part does not exhibit too much similarity and new part are introduced on a regular basis. In this case, benefits cannot be gained from Group Technology due to dissimilarity of parts. Because, new parts are regularly introduced, historical data does not have too much value to the process planner. However, aforementioned approach is a rough guideline for selecting the appropriate CAPP approach.

SAQ 1

- What is process planning?
- What are the various steps in developing a process plan?
- Why the need for CAPP arises?
- What are different approaches to CAPP? Describe briefly.
- Briefly describe the “Knowledge based Process Planning”.
- Write short notes on following:
    - Manual experience based process plan.
    - Computer Aided Process Plan.

Feature Recognition in CAPP

CAPP systems usually serve as link in integrating the CAD and CAM. However, it is only the partial link due to lack of part feature information provided by existing CAD/ Drafting system. Part feature information is an essential data for CAPP. In other words, it is a tedious job for CAPP to understand the three dimensional geometry of the designed part from CAD system in terms of their engineering meaning related to assembly and manufacturing. Generally, all CAPP planning method and systems suffered from such type of problem and is referred as feature recognition in CAPP.

Hence, objective of feature recognition is to bridge the gap between the database and automated process planning systems by automatically distinguishing the feature of a part from the geometry and topological data stored in the CAD system. The essence of feature recognition can easily be understood by taking an example as shown in figure. This figure is defined by a constructive solid geometry tree that represents a block primitive and a cylinder primitive combined by the Boolean operator "-". Shape and dimension can easily be identified by these schemes but, some higher level information is not provided by this scheme such as, whether the whole is blind hole or through hole. Such types of information are called as feature. Hence, features play a vital role in CAPP. In order to identify features and to solve CAD/CAPP interface problem, feature recognition is one of the most efficient technique.

Feature recognition transforms a general CAD model into an application specific feature model. In general, a generic part feature recognition system must be able to resolve following issues:

- Extract design information of a part.
- Identify all surfaces of part.
- Recognize reasons about\and\or interpret these surfaces in terms of Part features.

Once the features are classified, the automated planning system could develop the required process plan to make the part and hence, eliminate the need for a human to translate the CAD data into something that process planning system can understand.

Here, it is pertinent to mention that feature recognition is not only applicable to CAPP system but it can also be applied to various other engineering applications that require information about feature of parts classification and automated coding in GT.

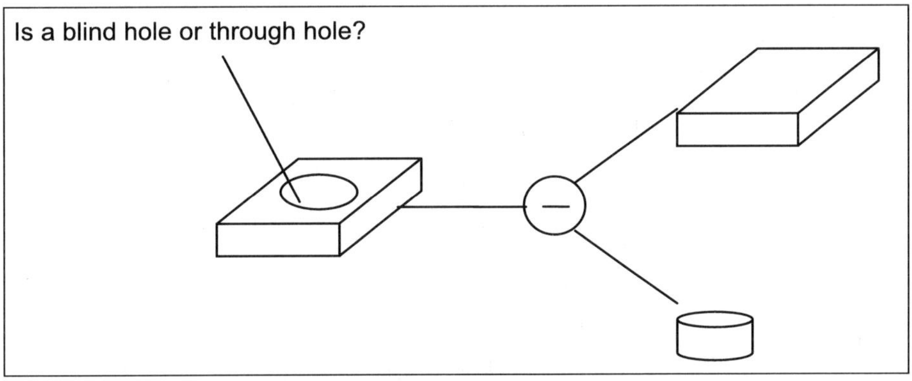

Advance Level Information.

Approaches to Part Feature Recognition

The most robust description of any part is provided through solid modelling. Therefore, these appear to be a logical starting point for developing a feature recognition system. Henderson and Anderson developed a system called FEATURES to perform automatic feature recognition using data from solid modelling system. A FEATURE simulates the human interpretation of part features. The system consists of a feature recognizer, extractor, and organizer. For objects containing swept features, this system performs well. This approach is encouraging because, conceptually, it can be applied for more complex parts. The FEATURES system uses the boundary representation (BREP) of a part, which denotes the faces, edges and vertices (FEV). Thus, features such as holes must be derived from more primitive data.

Representing FEV contained in a BREP graph. As a result a hole may be present as a collection of faces that must be recognized from the part data.

A number of approaches to part feature recognition for rotational as well as prismatic parts have been developed. These different approaches are enlisted as follows:

- Syntactic Pattern Recognition,
- State Transition Diagram and Automata,
- Geometry Decomposition Approach,
- Expert System Rule Logic,
- CSG (Set Theoretic) Approach,
- Graph Based Approaches.

The syntactic pattern recognition and/or expert logic approach are mainly applied for feature extraction of rotational part feature recognition. The complexity increases in case for prismatic parts due to lack of rotational property. In this case, the difficulty of both representation of a generic object and recognition of its feature increases extensively.

Attributed Adjacency Graph based Approach for Feature Recognition

Following three steps are involved in Graph based feature recognition:

- Generating graph based representation of the object to be recognized.
- Defining part features.
- Matching part features in the graph representation.

Generating Graph based Representation of the Object to be Recognized

During first step graphs are used for representation of the object. This step is necessary because data extracted from the data base are usually in the form of boundary representation (BREP) and cannot be used for feature recognition. Information regarding the type of face adjacency and relationship between the sets of faces should be expressed explicitly to recognize a feature. Here, attribute adjacency graph (AAG) has been used to demonstrate the recognition process.

Attributed Adjacency Graph

An AAG can be defined as a graph G = (N, A, T), where N is the set of nodes, A is the set of arcs, T is the set of attributes to arcs in A such that:

- For every face f in F, there exists a unique node n in N.
- For every edge e in E, there exists a unique arc a in A, connecting nodes n_i and n_j, correspond in to face f_i and face f_j, which share the common edge e.
- "t" is an attribute assigned to every arc a in A.
- t = 0 if the faces sharing the edge form a concave angle (or "inside" edge).
- t = 1 if the faces haring the edge form a convex angle (or "outside" edge).

The AAG is represented in the form of matrix as follows:

$$\begin{array}{c} \\ F_1 \\ F_2 \\ \vdots \\ \vdots \\ F_n \end{array}\begin{array}{c} \begin{array}{ccccc} F_1 & F_2 & \cdots & \cdots & F_n \end{array} \\ \begin{pmatrix} E_{1,1} & E_{1,2} & E_{1,3} & \cdots & E_{1,n} \\ \vdots & \vdots & \vdots & \vdots & \vdots \\ \vdots & \vdots & \vdots & \vdots & \vdots \\ \vdots & \vdots & \vdots & \vdots & \vdots \\ E_{n,1} & E_{n,2} & \cdots & \cdots & E_{n,n} \end{pmatrix} \end{array}$$

Where,

$$E_{ij} = \begin{cases} 0 & \text{if } F_i \text{ forms a concave angle with } F_j \\ 1 & \text{if } F_i \text{ forms a convex angle with } F_j \\ \phi & \text{if } F_i \text{ is not adjacent to } F_j \end{cases}$$

Hence it can easily be understood from above observation that AAG define the shape of a part uniquely up to its topology, if and only if the faces are cut orthogonally.

Definition of Part Feature

First we have to define what actually feature is? In general any shape can be feature if their manufacturing meanings are defined. There are mainly six features which are commonly used in manufacturing. These are step, slot, three side pocket, four side pocket, pocket (or blind hole) and through hole. Figures (a-f) represent some of the features and their surfaces are labeled.

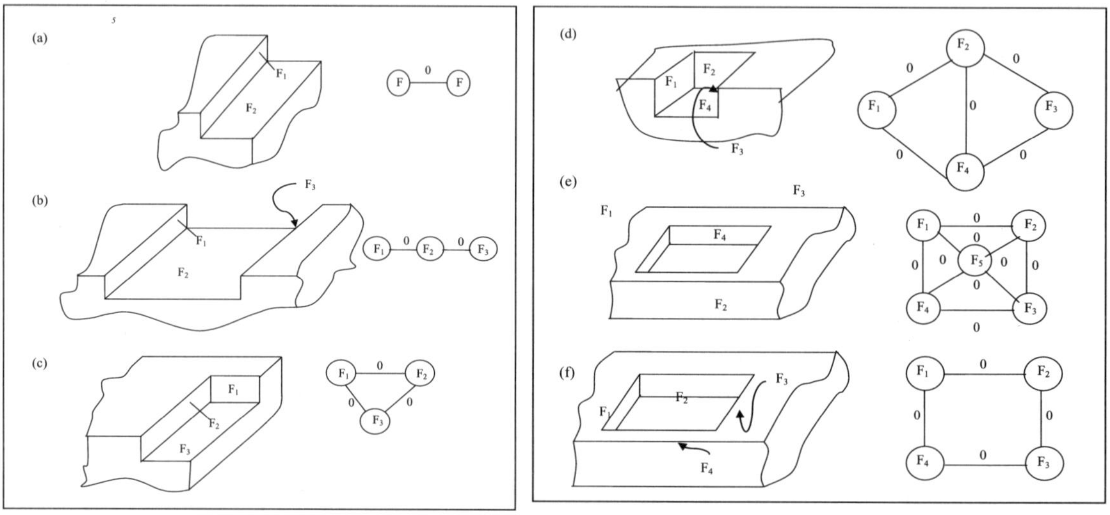

Various Features.

Four-side Pocket:

- F1 is adjacent to F2 and F4.
- F3 is adjacent to F2 and F4.
- F2 is adjacent to F1, F3, and F4.
- F4 is adjacent to F1, F3 and F2.
- F1 forms concave (90°) angles with F2 and F4.
- F2 forms concave (90°) angles with F3, F1 and F4.
- F3 forms concave (90°) angles with F2 and F4.
- F4 forms concave (90°) angles with F1, F2 and F3.

Blind Hole (Pocket):

- F1 is adjacent to F2, F4 and F5.
- F2 is adjacent to F1, F3 and F5.

- F3 is adjacent to F2, F4, and F5.
- F4 is adjacent to F1, F3, and F5.
- F5 is adjacent to all other surfaces of the pocket.
- F1 forms concave (90°) angles with F2, F4 and F5.
- F2 forms concave (90°) angles with F3, F1 and F5.
- F3 forms concave (90°) angles with F2 and F4 and F5.
- F4 forms concave (90°) angles F3, F1 and F5.
- F5 forms concave angle (90°) with all other surfaces of the pocket.

Matching the Features

After execution of step 1 and 2, the main task is to recognize AAG subgraph instead of recognizing machining features in parts. Because, complete AAG graph represents the part and subgraphs represent the features. It is computationally complex task to identify subgraphs in a graph. Although, there is no general way or algorithm to solve the problem.

Joshi and Chang, 1988 used an algorithm to identify the components of the graph that could form a feature. The algorithm is based on the following observations:

A face that is adjacent to all its neighbouring faces with a convex angle 270° does not form part of a feature. This observation is used as a basis for separating the original graph into subgraphs that could corresponds to features. The separation is done by deleting some nodes of the graph. The delete node rule follow IF THEN rule and given as follows. For all nodes:

- If (all incident arcs of a node have attributed "1").
- Then (delete this node (and all the incident arcs at the node) from AAG).

Delete node rule actually delete rows and column that represent the nodes in the matrix. Now feature recognition rule are applied to identify that whether or not a subgraph represent a feature. On the basis of feature definition some rules are written. In order to identify features some rules are given as follows:

For a Given AAG:

- It is a four side feature if: (The AAG subgraph has four nodes) and (The number of arcs with attribute "0" is the number of nodes plus one).
- It is a pocket features if: (The AAG subgraphs have five nodes) and (The number of arcs with attribute "0" is the number of attributes plus three).

In some of the cases, delete node rule fails. This is due to the intersection of features. Delete node rule cannot separate complete AAG graph into subgraphs, particularly for the case when features intersects. Hence, only for those cases when features are disjoint, delete node rule can be applied.

Following procedures are used to separate the subgraphs for intersecting features:

- Delete all 1 arcs.
- Form the subgraphs that may or may not represent the features.
- If not all the subgraphs, represent features, restore the one arc deleted within the subgraph.

The main task of this procedure is to separate the graphs into subgraphs. If the procedure is successful, it greatly reduces the computational effort otherwise graph remains unchanged.

An Illustrative Example

An example is taken from Li, which is illustrated as follows:

Example: It is easy for human to recognize that the part shown in figure has a slot and a pocket feature. In this example, however, we simulate the computer to apply the feature recognition algorithm discussed earlier. We, therefore, want to solve the following problem:

- Develop the AAG of the object,
- Give the matrix representation of the AAG,
- Recognize the features in this object.

Solution:

- First, we have to label each surface of the part. Given that the part is labeled as shown in figure, we develop the AAG as shown in figure from the definition of AAG. By the definition of AAG we mean that each surface of this part is represented by a node and each edge by an arc with attribute 1 or 0.
- For propose of inputting the AAG graph into the computer, we have to convert the graph to matrix form. The matrix representation of AAG is given as follows:
- Apply the delete node rule of the algorithm; that is delete the nodes connected with all "1" attributer arcs. Also delete these "1" arcs. Doing this on the adjacent matrix, we remove the rows and columns without 0 elements. For example, row 15 and column 15 represent this type of arc and can be deleted. After all such arcs are deleted; the present matrix will result into two unrelated sub-matrices.

Each sub-matrix represents a subgraph. We find that "0" in column 12, which is F12. We find that two disjoint subgraphs are generated. Figure gives the same structure as the one shown in figure, which corresponds to the slot feature. Thus the first feature that is recognized is the slot feature. In the computer this matching is achieved by applying the identifying rules. That is, if AAG subgroup has three nodes and the number of arcs with attribute "0" is 2, and then it is a slot. Also, because figure has the same structure as the one shown is figure, it is a pocket. As there are no more features to be recognized. We conclude that there are two features in this part, a slot and a pocket.

	F_1	F_2	F_3	F_4	F_5	F_6	F_7	F_8	F_9	F_{10}	F_{11}	F_{12}	F_{13}	F_{14}	F_{15}
F_1	9	0	9	0	0	1	9	9	9	9	9	9	9	9	9
F_2	.	9	0	9	0	1	9	9	9	9	9	9	9	9	9
F_3	.	.	9	0	0	1	9	9	9	9	9	9	9	9	9
F_4	.	.	.	9	9	1	9	9	9	9	9	9	9	9	9
F_5	.	.	.	.	9	9	9	9	9	9	9	9	9	9	9
F_6	.	.	.	.	.	9	1	9	9	9	9	9	1	1	1
F_7	.	.	.	.	.	.	9	1	9	9	9	9	9	1	1
F_8	.	.	.	.	.	.	.	9	1	9	9	9	9	1	1
F_9	.	.	.	.	.	.	.	.	9	1	9	9	9	1	1
F_{10}	.	.	.	.	.	.	.	.	.	9	1	9	9	1	1
F_{11}	s	y	m	m	e	t	r	y	.	.	9	0	9	1	1
F_{12}	.	.	.	.	.	.	.	.	.	.	.	9	0	1	1
F_{13}	.	.	.	.	.	.	.	.	.	.	.	.	9	1	1
F_{14}	.	.	.	.	.	.	.	.	.	.	.	.	.	9	9
F_{15}	.	.	.	.	.	.	.	.	.	.	.	.	.	.	9

Computer-aided Software Engineering

Computer-aided software engineering (CASE) is the implementation of computer facilitated tools and methods in software development. CASE is used to ensure a high-quality and defect-free software. CASE ensures a check-pointed and disciplined approach and helps designers, developers, testers, managers and others to see the project milestones during development.

CASE can also help as a warehouse for documents related to projects, like business plans, requirements and design specifications. One of the major advantages of using CASE is the delivery of the final product, which is more likely to meet real-world requirements as it ensures that customers remain part of the process.

CASE illustrates a wide set of labor-saving tools that are used in software development. It generates a framework for organizing projects and to be helpful in enhancing productivity. There was more interest in the concept of CASE tools years ago, but less so today, as the tools have morphed into different functions, often in reaction to software developer needs. The concept of CASE also received a heavy dose of criticism after its release.

CASE Tools

The essential idea of CASE tools is that in-built programs can help to analyze developing systems in order to enhance quality and provide better outcomes. Throughout the 1990, CASE tool became part of the software lexicon, and big companies like IBM were using these kinds of tools to help create software.

Various tools are incorporated in CASE and are called CASE tools, which are used to support different stages and milestones in a software development life cycle.

Types of CASE Tools

- Diagramming Tools: It helps in diagrammatic and graphical representations of the data and system processes. It represents system elements, control flow and data flow among different software components and system structure in a pictorial form. For example, Flow Chart Maker tool for making state-of-the-art flowcharts.
- Computer Display and Report Generators: It helps in understanding the data requirements and the relationships involved.
- Analysis Tools: It focuses on inconsistent, incorrect specifications involved in the diagram and data flow. It helps in collecting requirements; automatically check for any irregularity, imprecision in the diagrams, data redundancies or erroneous omissions. For example:
    - Accept 360, Accompa, CaseComplete for requirement analysis.
    - Visible Analyst for total analysis.
- Central Repository: It provides the single point of storage for data diagrams, reports and documents related to project management.
- Documentation Generators: It helps in generating user and technical documentation as per standards. It creates documents for technical users and end users. For example, Doxygen, DrExplain, Adobe RoboHelp for documentation.
- Code Generators: It aids in the auto generation of code, including definitions, with the help of the designs, documents and diagrams.

Advantages of the CASE Approach

- As special emphasis is placed on redesign as well as testing, the servicing cost of a product over its expected lifetime is considerably reduced.

- The overall quality of the product is improved as an organized approach is undertaken during the process of development.
- Chances to meet real-world requirements are more likely and easier with a computer-aided software engineering approach.
- CASE indirectly provides an organization with a competitive advantage by helping ensure the development of high-quality products.

References

- Computer-aided-process-planning-a-state-of-art-1825014: academia.edu, Retrieved 28 March, 2019
- Computer-aided-quality-assurance-caqc: jensen-consulting.co.uk, Retrieved 23 April, 2019
- Computer-aided-quality-control, cad-autocad-reviews-tips-64482: brighthubengineering.com, Retrieved 24 February, 2019
- Computer-aided-manufacturing-cam- 4698: techopedia.com, Retrieved 17 May, 2019
- Introduction-to-cad, knowledge-base: designtechcadacademy.com, Retrieved 18 January, 2019
- Computer-aided-process-planning-CAPP: searcherp.techtarget.com, Retrieved 14 July, 2019
- Computer-aided-software-engineering-case: geeksforgeeks.org, Retrieved 25 August, 2019

3

Industrial Process Control Systems

Industrial control systems deal with the operation of control systems and instrumentation used in industrial processes. It includes discrete process control, computer process control, computer numerical control, motion control system, supervisory control, artificial neural network, etc. The topics elaborated in this chapter will help in gaining a better perspective of these industrial control systems.

Industrial control system (ICS) is a collective term used to describe different types of control systems and associated instrumentation, which include the devices, systems, networks, and controls used to operate and automate industrial processes. Depending on the industry, each ICS functions differently and are built to electronically manage tasks efficiently. Today the devices and protocols used in an ICS are used in nearly every industrial sector and critical infrastructure such as the manufacturing, transportation, energy, and water treatment industries.

There are several types of ICSs, the most common of which are Supervisory Control and Data Acquisition (SCADA) systems, and Distributed Control Systems (DCS). Local operations are often controlled by so-called Field Devices that receive supervisory commands from remote stations.

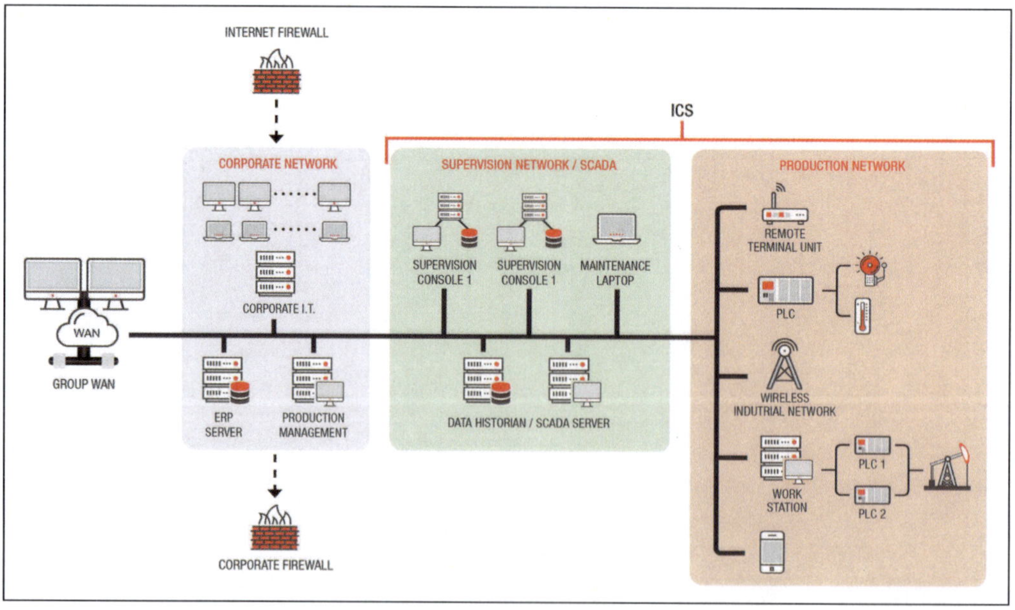

Common ICS Threats

Securing industrial systems is no trivial task. Most were built before the first cyber threat surfaced, and did not have built-in external security controls factored into their design. Understanding some of the most common industrial control system threats is the first step that any industrial organization can take to protect their network.

External Threats and Targeted Attacks

When you consider that industrial control systems often come under the umbrella of chemical engineering, manufacturing, distribution and healthcare, it is little wonder that these systems are often targeted by terrorist groups, hacktivists and other groups with malicious intent. The aim of politically motivated attacks is usually focused on causing physical damage or operational disruption, while industrial espionage attacks will be more focused on stealing or damaging Intellectual Property (IP).

Internal Threats

Insider threats are well documented when it comes to IT networks, but they can also pose a huge risk to industrial networks. From disgruntled employees to contractors with an ax to grind, the internal threat is real. Most ICS networks require little to no authentication or encryption that controls or restricts user activity. This means that any insider will generally have unlimited access to any device that exists on the network, including SCADA applications and other critical components. Systems that have been upgraded to connect to a digital interface can be easily compromised by malware or a USB device used to download sensitive data.

Human Error

It is human nature to make mistakes. However, when mistakes are made on an industrial control systems network they can be costly and have a huge impact on operations and reputation. In fact, in many cases, human error is considered the biggest threat to an ICS network. Human errors can include making incorrect configurations, PLC programming errors or forgetting to monitor key metrics or alerts.

Discrete Process Control

Discrete control system (DCS) is a system that is used to control production systems that are found in one location. In a DCS, a set point is sent to the controller that is capable of instructing valves, or even an actuator, to operate in such a way that the desired set point is maintained. Data from the field can be stored for future reference, used for simple process control, or even used for advanced control strategies with data from another part of the plant. Each DCS uses a centralized supervisory control loop to manage multiple local controllers or devices that are part of the overall production process. This gives industries the ability to quickly access production and operation data. And by using multiple devices within the production process, a DCS is able to reduce the impact of a single fault on the overall system.

Discrete control is one of the major subsystems in the industrial automation and needs of the manufacturing sectors are the primary driver for the growth and maturity. The discrete control and automation have started the journey with the sensors and relays wired to realize the logic. However, the flexibility expected by the manufacturing industry brought forward the concepts such as PLC fuelled by the microprocessors and microcontrollers and matured as emergency shutdown systems (ESDs) by adopting networking and distributed processing technologies.

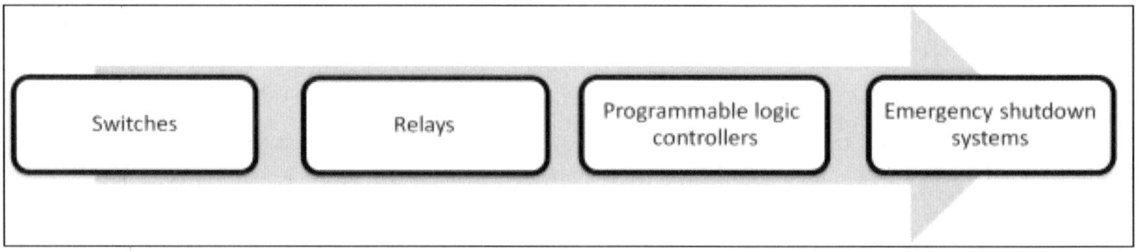

Switches

In the past, machine control could be achieved by wiring switches and relays physically together to perform logic functions. Hard-wired ladder diagrams were created for the circuit to work; current had to flow through each rung. The main drawback of this system was that a change in logic function required a physical rewire of the circuit. This involved maintenance personnel conducting a time-consuming job and often resulted in lost production time, as the machine would have to be switched off.

Relay

Relays serve as switching, timing, and multiplying mechanisms for input devices, such as push buttons, selector switches, and photoelectric sensors. Relays are fairly intuitive; however, as mechanical devices, they do not offer the same programming and troubleshooting flexibility found in modern programmable control systems. Relays are also known for taking up considerable amounts of space, requiring extensive wiring, and needing regular maintenance.

PLC

The first programmable controller, introduced in 1970, was developed in response to a demand from General Motors for a solid-state system that had the flexibility of a computer, yet could be programmed and maintained by plant engineers and technicians.

These early programmable controllers took up less space than the relays, counters, timers, and other control components they replaced, and they offered much greater flexibility in terms of their reprogramming capability. The initial programming language, based on the ladder diagrams and electrical symbols commonly used by electricians, was key to industry acceptance of the programmable controller. Because programmable controllers can be programmed in relay ladder logic, it is relatively simple to convert electrical diagrams to the programmable controller program. This process involves defining the rules of operation for each control point, converting these rules to ladder logic, and identifying and labeling outputs (addressing). Today's work force has a mix of engineers – some of whom have been around for a while and are familiar with ladder logic as well as newer engineers more comfortable with computer-centric based programming and control. This has led to a mix of programming technologies that are applied based on user background and application need.

ESD

The ESD shall minimize the consequences of emergency situations, related to typically uncontrolled flooding, escape of hydrocarbons, or outbreak of fire in hydrocarbon-carrying areas or areas that may otherwise be hazardous. Traditionally risk analyses have concluded that the ESD system is in need of a high safety integrity level (SIL), typically SIL 2 or 3. Basically the system consists of field-mounted sensors, valves and trip relays, system logic for processing of incoming signals, alarm, and HMI units. The system is able to process input signals and activating outputs in accordance with the cause and effect charts defined for the installation. Shutdown of part systems and equipment is to isolate hydrocarbon inventories, isolate electrical equipment, prevent escalation of events, stop hydrocarbon flow, depressurize/blow down, emergency ventilation control, and close watertight doors and fire doors.

Computer Process Control

The use of digital computers to control industrial processes had its origins in the continuous process industries in the late 1950s. Prior to then, analog controllers were used to implement continuous control, and relay systems were used to implement discrete control. At that time, computer technology was in its infancy, and the only computers available for process control were large, expensive mainframes. Compared with today's technology, the digital computers of the 195Ch; were slow, unreliable, and not well suited to process control applications. The computers that were installed sometimes cost more than the processes they controlled. Around 1960, digital computers started replacing analog controllers in continuous process control applications; and around 1970, programmable logic controllers started replacing relay hanks in discrete control applications. Advances III computer technology since the 1960s and 1970s have resulted in the development of the microprocessor. Today, virtually all industrial processes, certainly new installations, are controlled by digital computers based on microprocessor technology.

Control Requirements

Whether the application involves continuous control, discrete control or both, there are certain basic requirements that tend to be common to nearly all process control applications. By and large, they are concerned with the need to communicate and interact with the process on a real time basis. A real time controller is able to respond to the process within a short enough time periods that process performance is not degraded. Factors that determine whether a computer controller can operate in real time include: (1) the speed of the controller's central processing unit (CPU) and its interfaces, (2) the controller's operating system, (3) the design of the application software, and (4) the number of different input/output events to which the controller is designed to respond. Real time control usually requires the controller to be capable of multitasking, which means coping with multiple tasks concurrently without the tasks interfering with one another.

There are two basic requirements that must be managed by the controller to achieve real time control:

- Process initiated interrupts: The controller must be able to respond to incoming signals from the process. Depending on the relative importance of the signals, the computer may need to interrupt execution of a current program to service a higher priority need of the process. A process initiated interrupt is often triggered by abnormal operating conditions, indicating that some corrective action must be taken promptly.
- Timer initialed actions: The controller must be capable of executing certain actions at specified points in time. Timer initiated actions can be generated at regular time intervals, ranging from very low values to several minutes or they can be generated at distinct points in time. Typical timer initiated actions in process control include: (1) scanning sensor values from the process at regular sampling intervals, (2) turning on and off switches, motors, and other binary devices associated with the process at discrete points in time during the work cycle, (3) displaying performance data on the operator's console at regular times during a production run, and (4) re-computing optimal process parameter values at specified times.

Capabilities of Computer Control

The above requirements can he satisfied by providing the controller with certain capabilities that allow it 10 interact on a real time basis with the process and the operator. The capabilities are: (1) polling, (2) interlocks. (3) interrupts system, and (4) exception handling.

Polling (Data Sampling)

In computer process control, refers 10 the periodic sampling of data that indicates the status of the process. When the data consist of a continuous analog signal sampling means that the continuous signals substituted with a series of numerical values that represent the continuous signal at discrete moments in time. The same kind of substitution holds for discrete data except that the number of possible numerical values the data can take on more limited certainly the case with binary data other names used for polling include sampling and scanning.

In some polling procedure simply requests whether any changes have the last polling cycle and then collects only the new data from shorten the cycle time required for polling. Issues related to:

- Polling order: The polling order is the sequence in which the different data collection points of the process are sampled.
- Polling formal: This refers to the manner in which the sampling procedure is designed. The alternatives include: (a) entering all new data from all sensors and other devices very polling cycle; (b) updating the control system only with data that have changed since the last polling cycle; or (c) using highlevel scanning, in which only certain key data are normally collected each polling cycle (highlevel scanning), but if the data indicates some irregularity in the process. A low level scan undertaken to collect more complete data to ascertain the source of the irregularity.
- Interlocks: An interlock is a safeguard mechanism for coordinating the activities of two or more devices and preventing other device from interfering with the others. In process control interlocks provide a means by which the controller is able to sequence the activities in a work cell, ensuring that the actions of one piece of equipment arc completed before the next piece of equipment begins its activity. Interlocks work by regulating the flow of control signals back and forth between the controller and the external devices.

There are two types of interlocks, input interlocks and output interlocks, where input and output are defined relative to the controller. An input interlock is a signal that originates from an external device (e.g., a limit switch, sensor, or production machine) and is sent to the controller. Input interlocks can be used for either of the following functions:

- To proceed with the execution of the work cycle program. For example, the production machine communicates a signal to the controller that it has completed its processing of the part. This signal constitutes an input interlock indicating that the controller can now proceed to the next step in the work cycle, which is to unload the part.

- To interrupt the execution of the work cycle program. For example, while unloading the part from the machine, the robot accidentally drops the part. The sensor in its gripper transmits an interlock signal to the controller indicating that the regular work cycle sequence should be interrupted until corrective action is taken.

An output interlock is a signal sent from the controller to some external device. It is used to control the activities of each external device and to coordinate its operation with that of the other equipment in the cell. For example, an output interlock can be used to send a control signal to a production machine to begin its automatic cycle after the work part has been loaded into it.

Interrupt System

Closely related to interlock is the interrupt system. As suggested by our discussion of input interlocks, there are occasions when it becomes necessary for the process or operator to interrupt the regular controller operation to deal with more pressing matters. All computer systems are capable of being interrupted; if nothing else, by turning off the power. A more sophisticated interrupt system is required for process control applications. An interrupt system is a computer control feature that permits the execution of the current program to be suspended to execute another program or subroutine in response to an incoming signal indicating a higher priority event. Upon receipt of an interrupt signal, the computer system transfers to a predetermined subroutine designed to deal with the specific interrupt. The status of the current program is remembered so that its execution can be resumed when servicing of the interrupt has been completed.

Interrupt conditions can be classified as internal or external internal interrupts are generated by the computer system itself. These include timer initiated events, such as polling of data from sensors connected to the process, or sending commands to the process at specific points 10 clock time. System and program initiated interrupts are also classified as internal because they are generated within the system. External interrupts are external to the computer system; they include process-initiated interrupts and operator inputs.

Table: Possible priority levels in an Interrupt system.

Priority Level	Computer Function
1 (lowest priority)	Most operator inputs
2	System and program interrupts
3	Timer interrupts
4	Commands to process
5	Process interrupts
6 (highest priority)	Emergency stop (operator input)

An interrupt system is required in process control because it is essential that more important programs (ones with higher priority) be executed before less-important programs (ones with lower priority). The system designer must decide what level of priority should be attached to each control function. A function at a given priority level cannot interrupt a function at the same priority level. The number of priority levels and the relative importance of the functions depend on the requirement of the individual process control situation. For example, emergency shutdown of a process because of safety hazards would occupy a very high priority level, even though it may be an operator initiated interrupt. Most operator inputs would have low priorities.

The system designer must decide what level of priority should be attached to each control function. A higher priority function can interrupt a lower primary function. A function at a given priority level cannot interrupt a function at the same priority level. For example, emergency shutdown of a process because of safety hazards would occupy a very high priority level, even though it may be an operator initiated interrupt. Most operators in" puts would have low priorities.

One possible organization of priority rankings for process control functions is shown table. Of course, the priority system may have more or less than the number of levels shown here depending on the control situation. For example, some process interrupts may be more important than others and some system interrupts may take precedence over certain process interrupts, thus requiring more than the six levels indicated in table.

To respond 10 the various levels of priority defined for a given control application, an interrupt system can have one or more interrupt levels. A single level interrupt system has only two modes of operation: normal mode and interrupt mode. The normal mode can be interrupted, but the interrupt mode cannot. This means that overlapping interrupts arc serviced on a first come. First served basis, which could have potentially hazardous consequences it an important process interrupt was forced to wait its tum while a series of less important operator and system interrupts were serviced. A multilevel interrupt system has a normal operating mode plus more than one interrupt level. The normal mode can be interrupted why any interrupt level. But the interrupt levels have relative priorities that determine which functions can interrupt others. Example illustrates the difference between the single level and multilevel interrupt systems.

Example: Single Level versus Multilevel Interrupt Systems.

Three Interrupts representing tasks of three different priority levels arrive for service in the reverse order of (their respective priorities. Task 1 with the lowest priority, arrives first. Shortly later higher priority Task 2 arrives. And shortly later, highest priority Task 3 arrives. How would the computer control system respond under (a) a single level interrupt system and (b) a multilevel interrupt system?

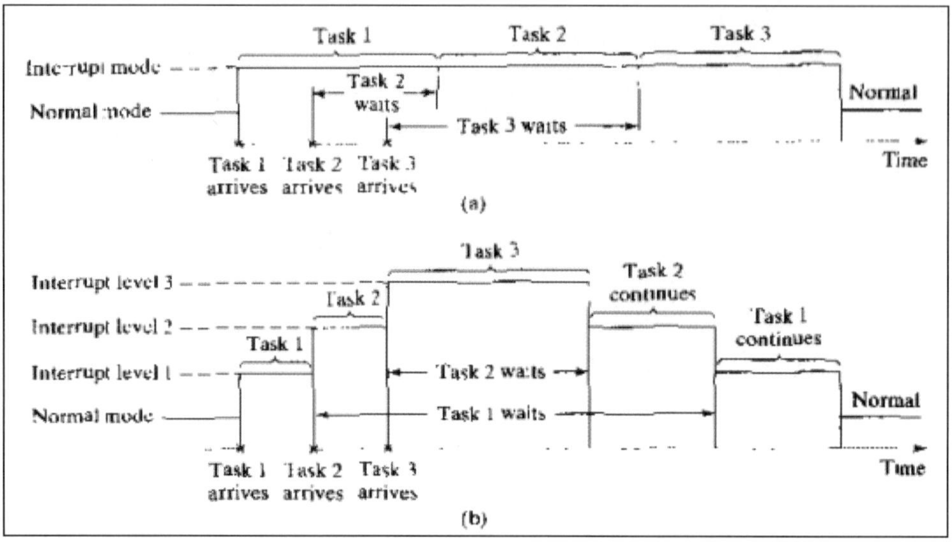

Response of the computer control system in example to the three different priorities interrupts for (a) a single-level interrupt system and (b) a multilevel interrupt system. Task 3 has the highest level priority. Task 1 has the lowest level. Tasks arrive for servicing in the order 1, then 2, then 3. In (a), Task 3 must wait until Tasks 1 and 2 have been completed. In (b), Task 3 interrupts execution of Task 2, whose priority level allowed it to interrupt Task 1.

Exception Handling

In process control, an exception is an event that is outside the normal or desired operation of the process or control system. Dealing with the exception is an essential function in industrial process control and generally occupies a major portion of the control algorithm. The need for exception handling may be indicated through the normal polling procedure or by the interrupt system. Examples of events that may invoke exception handling routines include:

- Product quality problem.
- Process variables operating outside their normal ranges.
- Shortage of raw materials or supplies necessary to sustain the process.
- Hazardous conditions such as a fire.
- Controller malfunction.

Forms of Computer Process Control

There are various ways in which computers can be used to control a process. First, we can distinguish between process monitoring and process control as illustrated in figure.

In process monitoring the computer is used to simply collect data from the process, while in process control, the computer regulates the process. In some process control implementations, certain actions are implemented by the control computer that requires no feedback data to be collected from the process. This is open loop control. However, in most cases; some form of feedback or interlocking is required to ensure that the control instructions have been properly carried out. This more common situation is closed loop control.

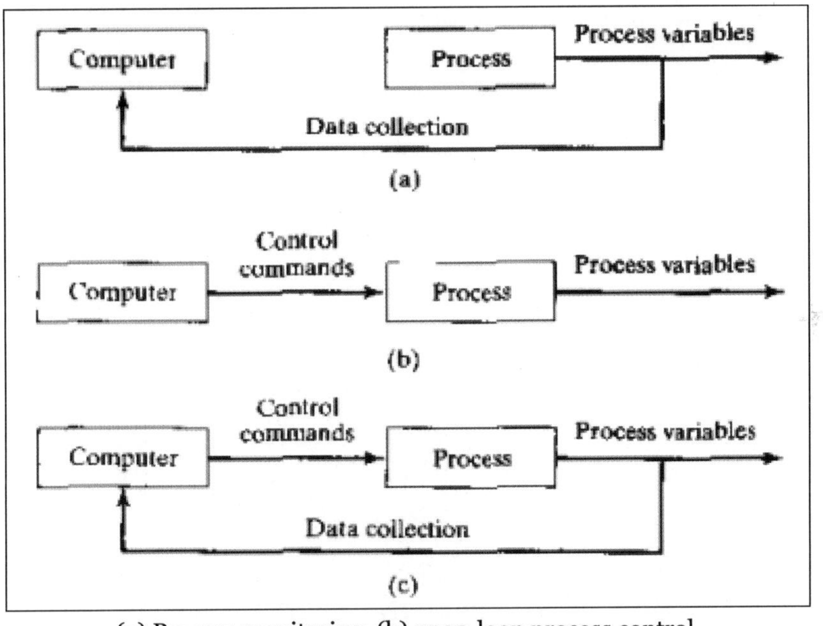

(a) Process monitoring, (b) open-loop process control, and (c) closed-loop process control.

Here, we survey the various forms of computer process monitoring and control, all huts one of which commonly used in industry today. The survey covers the following categories. (1) computer process monitoring. (2) direct digital control. (3) numerical control and robotics. (4) programmable logic controllers) supervisory control, and (6) distributed control systems and personal computers. The second category, direct digital control, represents a transitory phase in the evolution of computer control technology. In its pure OM, it is no longer used today. However we briefly describe DDC to expose the opportunities it contributed.

Computer Process Monitoring

Computer process monitoring is one of the ways in which the computer can be interfaced with a process. Computer process monitoring involves the use of the computer to observe the process and associated equipment and to collect and record data from the operation. The computer is not used to directly control the process. Control remains in the hands of humans who use the data to guide them in managing and operating the process.

The data collected by the computer in computer process monitoring can generally be Classification into three categories:

- Process Data: These are measured values of input parameters and output variable that indicate process performance. When the values are found to indicate a problem, the human operator takes corrective action.
- Equipment Data: These data indicate the status of the equipment in the work cell. Functions served by the data include monitoring machine utilization, scheduling tool changes, avoiding machine breakdowns. Diagnosing equipment malfunctions, and planning preventive maintenance.
- Product Data: Government regulations require certain manufacturing industries to collect and preserve production on their products. The pharmaceutical and medical supply industries are prime examples. Computer monitoring is the most convenient, means of satisfying these regulations. A firm may also want 10 collect product data for its own use.
- Collecting Data: Collecting data from factory operations can be accomplished by any of several means Shop data can be entered by workers through manual terminals located throughout the plant or can be collected automatically by means of limit switches, sensor systems, bar code readers, or other devices.

Direct Digital Control

Direct digital control was certainly one of the important steps in the development of computer process control. Let us briefly examine this computer control mode and its limitations, which motivated improvement leading to modern computer control technology. Direct digital controller (DDC) IS a computer process control system in which certain components in a conventional analog control system are replaced by the digital computer. The regulation of the process is accomplished by the digital computer on a timeshared, sample data basis rather than by the many individual analog components working in a dedicated continuous manner. With DDC, the computer calculates the desired values of the input parameters and set points, and these values are applied through a direct link to the process; hence the name "direct digital" control.

The difference between direct digital control and analog control can be seen by comparing figures. The first figure shows the instrumentation for a typical analog control loop. The entire process would have many individual control loops, but only one is shown here. Typical hardware components of the analog control loop include the sensor and transducer, an instrument for displaying the output variable (such an instrument is not always included in the loop), some means for establishing the set point of the loop (shown us a dial in the figure, suggesting that the setting is determined by a human operator). A comparator (to compare set point with measured output variable), the analog controller amplifier and actuator that determines the input parameter to the process.

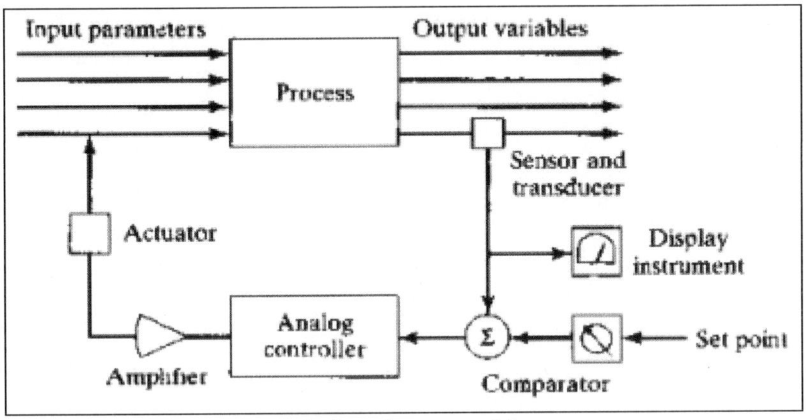

A typical analog control loop.

In the DDC system, some of the control loop components remain unchanged, including (probably) the sensor and transducer as well as the amplifier and actuator. Components likely to be replaced in DDC include the analog controller, recording and display instruments, set point dials, and comparator. New components in the loop include the digital computer, analog-to-digital and digital-to-analog converters (ADCs and DACs), and multiplexers share data from different control loops with the same computer.

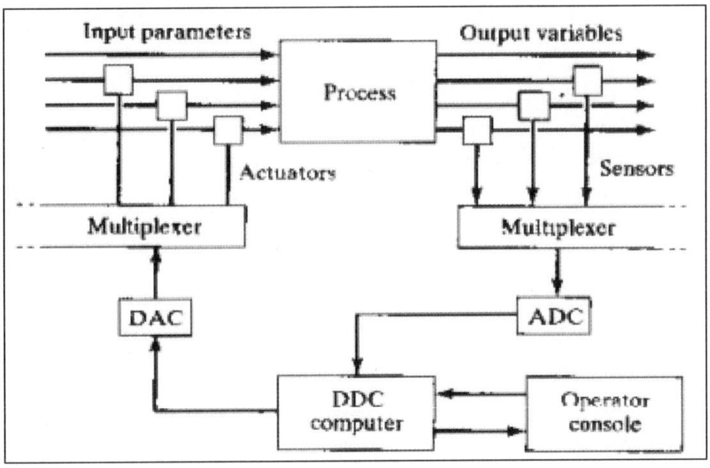

Components of a DDC system.

DDC was originally conceived as a more-efficient means of performing the same kinds of control actions as the analog components it replaced. However, the practice of simply using the digital computer to imitate the operation of analog controllers seems to have been a transitional phase in computer process control. Additional opportunities for the control computer were soon recognized, including:

- More control options than traditional analog: With digital computer control, it is possible to perform more complex control algorithms than with the conventional proportional integral derivative control modes used by analog controllers; for example on/off control or non linarites in the control functions can he implemented.

- Integration and optimization of multiple loops: This is the ability to integrate feedback measurements from multiple loops and to implement optimizing strategies to improve overall process performance.
- Adjust the control programmes: Using a digital computer makes it relatively easy to change the control algorithm if that becomes necessary by simply reprogramming the computer. Reprogramming the analog control loop is likely to require hardware changes that are more costly and less convenient.

These enhancements have rendered the original concept of direct digital control more or less obsolete. In addition, computer technology itself has progressed dramatically so that much smaller and less expensive yet more powerful computers are available for process control than the large mainframes available in the early 1960 and this has allowed computer process control to be economically justified for much smaller scale processes and equipment. It has also motivated the use of distributed control systems, in which a network of microcomputers is utilized to control a complex process consisting of multiple unit operations and machines.

Numerical Control and Robotics

Numerical control is another form or industrial computer control. It involves the use of the computer (again, a microcomputer) to direct a machine tool through a sequence of processing defined by a program of instructions that specifies the details of each step and their sequence. The distinctive feature of NC is control of the relative position of a tool with respect to the object (work part) being processed. Computations must be made to determine the trajectory that must be followed by the cutting tool to shape the part geometry.

Closely related to NC is industrial robotics, in which the joints of the manipulator (robot arm) are controlled to move the endofarm through a sequence of positions during the work cycle. As in NC the controller must perform calculations during the work cycle to implement motion interpolation. feedback control, and other functions in addition, a robotic work cell usually includes other equipment besides the robot, and the activities of the other equipment in the work cell must be coordinated with those of the robot. This coordination is achieved using interlocks.

Programmable logic controllers (FLCs) were introduced around 1970 as an improvement on the electromechanical relay controllers used at the time to implement discrete control in the discrete manufacturing industries. The evolution of PLCs has been facilitated by advances in computer technology, and present day PLCs are capable of much more than the 1970sera controllers. We can define a modern programmable logic controller as a microprocessor based controller that uses stored instructions in programmable memory to implement logic, sequencing, timing, counting, and arithmetic control functions for controlling machines and processes. Today's PLCs are used for both continuous control and discrete control applications in both the process industries and discrete manufacturing.

Supervisory Control

The term supervisory control is usually associated with the process industries but the concept applies equally well to discrete manufacturing automation where it corresponds 10 the cell or system level. Thus, supervisory control coincides closely with coordination control in the ANSIIISA·S88 Standard. Supervisory control represents a higher level of control than the preceding forms of process control in general these other types of control systems are interfaced directly to the process. By contrast supervisory control is often superimposed on these process level control systems and directs their operations. The relationship between supervisory control and the process level control techniques is illustrated in figure.

In the context of the process industries, supervisory control denotes a control system that manages the activities of a number of integrated unit operations to achieve certain economic objectives for the process. In some applications, supervisory control is not much more than regulatory control or feed forward control. In other applications, the supervisory control system is designed to implement optimal or adaptive control. It seeks to optimize some well-defined objective function which is usually based on economic criteria such as yield production rate, cost, quality, or other objectives that pertain to process performance.

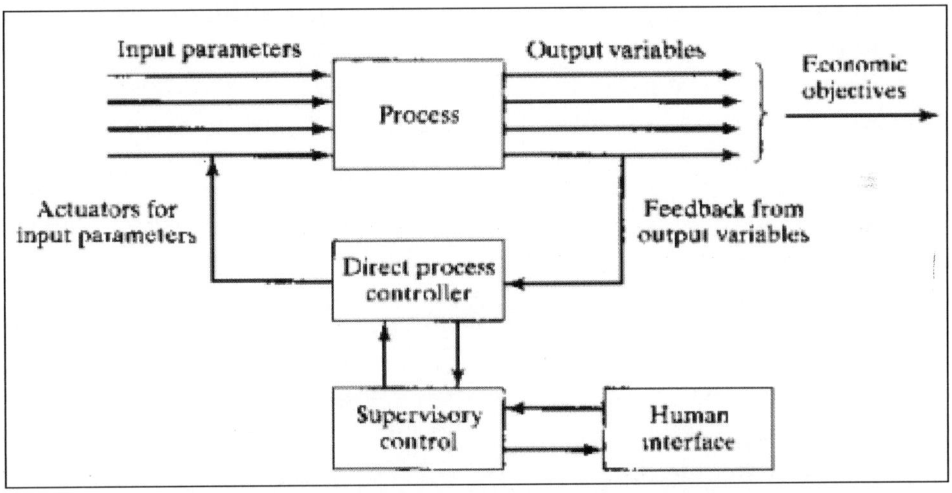

Distributed Control Systems and Personal Computers.

Development of the microprocessor has had a significant impact on the design of' control sysrcrns. We consider two related aspects of this impact: (I) distributed control systems and (2) the use of personal computers in control systems. Before discussing these topics let us provide a brief background of the microprocessor and its uses.

Microprocessors

A microprocessor is an integrated circuit chip containing the digital logic elements needed to perform arithmetic calculations, execute instructions stored in memory and carry out other data processing tasks. The digital logic elements and their interconnections in the circuit form a built in set of instructions that determines the run microprocessor. A very common function is to serve as the central processing unit microcomputer. By definition, a microcomputer is simply a small digital computer whose CPU is a microprocessor and which performs the functions of a computer. These basic functions consist of data manipulation and computation carried out according to software stored in memory to accomplish user applications. The most familiar and widely used example of a microcomputer is the persona computer (PC), usually programmed with software for business and personal applications.

With the development of the microprocessor, it became feasible to connect multiple microcomputers together to share and distribute the process control workload. The term distributed control system (DCS) is used 10 describe such a configuration which consists of the following components and features:

- Multiple process control systems located throughout the plant to control the individual loops and devices of the process.
- A central control room equipped with operator stations, where supervisory control of the plant is accomplished.
- Local operator stations distributed throughout the plant. This provides the DCS with redundancy. If a control failure occurs in the central control room. The local operator stations take over the central control functions. If a local operator station fails, the other local operator stations assume the functions of the failed station.

All process and operator stations interact with each other by means of a communications network, or data highway, as it is often called.

These component, arc illustrated in a typical configuration of a distributed process control system presented in figure. There are a number of benefits and advantages of the DCS (1) A DCS can be installed for a given application in a very basic configuration, then enhanced and expanded as needed in the future; (2) since the system consists of multiple computers, this facilitates parallel multitasking; (3) because of its multiple computers, a DCS has built in redundancy; (4) control cabling is reduced compared with a central computer control configuration.

Development of DCS, started around 1970. One of the first commercial systems was Honeywell's TDC 2000, introduced in 1975. The first DCS applications were in the process industries. In the discrete manufacturing industries, programmable logic controllers were introduced about the same time. The concept of distributed control applies equally well to PLCs; that is, multiple PLCs located throughout a factory to control individual pieces of equipment hut integrated by means of a common communications network. Introduction of the PC shortly after the Des and PLC, and LIS subsequent increase in computing power and reduction in cost over the years, have stimulated a significant growth in the adoption of PC based DCSs for process control applications.

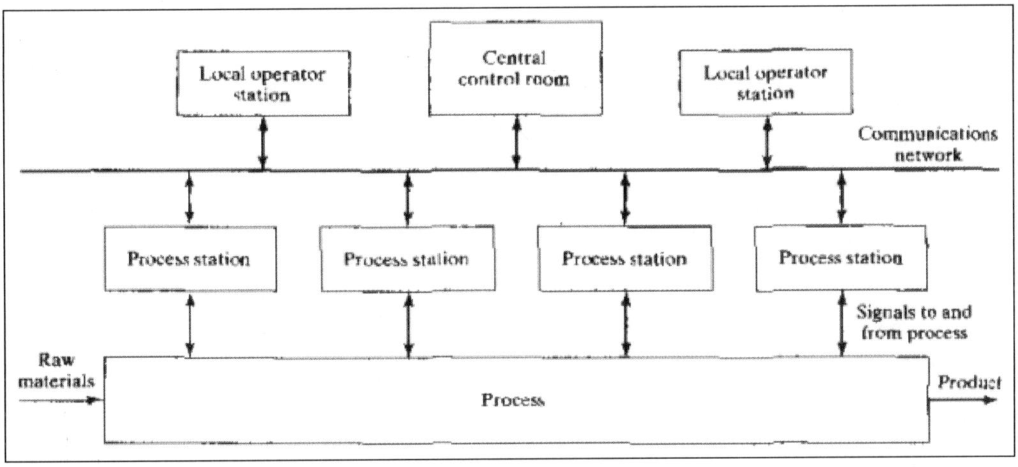

Distributed control system.

Today, pcs dominate the computer world. They have become the standard 1001 by which business is conducted whether in manufacturing or in the service sector. Thus, it is no surprise that PCs are being used in growing numbers in process control applications. Two basic categories of PC applications in process control can be distinguished: (1) operator interface and (2) direct control. Whether used as the operator interface or for direct control. PCs are likely to be networked with other computers to create DCSs.

When used as the operator interface, the PC is interfaced to one or more PLCs or other devices (possibly other microcomputers) that directly control the process. Personal computers have been used to perform the operator interface function since the early 1980s. In this function, the computer performs certain monitoring and supervisory control functions, hut it does not directly control the process. Advantages of using a PC as only the operator interface include: (1) the PC' provides a user friendly interface for the operator; (2) the PC can be used for all of the conventional computing and data processing functions that PCs traditionally perform; (3) the PLC or other device that is directly controlling the process is isolated from the Pc so a PC failure will not disrupt control of the process; and (4) the computer can be easily upgraded as PC technology advances and capabilities improve. While the PLC control software and connections with the process can remain in place.

Direct control means that is interfaced directly to the process and controls its operations in real time. The traditional thinking has been that it is too risky to permit the PC to directly control the production operation. If the computer were to fail, the uncontrolled operation might stop working, produce a defective product, or become unsafe. Conventional PCs, equipped with the usual business oriented operating system and applications software, are designed for computing and data processing functions, not for process control. They are not intended to be interfaced with an external process in the manner necessary for real time process control. Finally, most PCs are designed to be used in an office environment, not in the harsh factory atmosphere.

Recent advances in both PC technology and available software have challenged this traditional thinking. Starting in the early 1990s, PCs have been installed at an accelerating pace for direct control of industrial processes. Several factors can be identified that have enabled this trend:

- Widespread similarity with PCs.
- Availability of high performance PCs.
- Trend toward open architecture philosophy in control systems design.

Microsoft's Windows NTH' (the latest version is Windows 2000IM) as the operating system of choice.

The PCs widely known to the general population in the United States and other industrialized nations. A large and growing number of individuals own them. Many others who do not personally own them use them at work, User friendly software for the home and business has certainly contributed to the popularity of PCs. There is a growing expectation by workers that they be provided with a computer in their workplace.

Enterprise Wide Integration of Factory Data

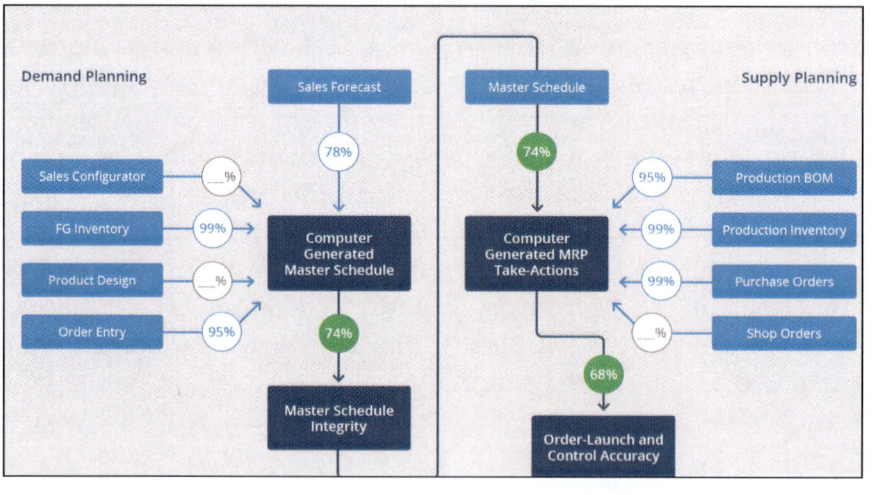

The most recent progression in PC based distributed control is enterprise wide integration of factory operations data. This is a trend that is consistent with modem information management and worker empowerment philosophies. These philosophies assume fewer levels of company management and greater responsibilities for frontline workers in sales, order scheduling and Production. The networking technologies that allow such integration are available. Windows 20001 provides a number of built-in and optional features for connecting the industrial control system in the factory to enterprise wide business systems and supporting data exchange between various applications (e.g. allowing data collected to be used in analysis packages. such as Excel spreadsheets). Following are some of the capabilities that are enabled by making process data available throughout the enterprise sales personnel can provide realistic estimates on delivery dates to customers, based on current shop loading.

Order trackers are able to provide inquiring customers with current status information on their orders.

Quality control personnel are made aware of real or potential quality problems on current orders, based on access to quality performance histories from previous orders. Cost accounting has access to the most recent production cost data.

Production personnel can access part and product design details to clarify ambiguities and do their job more effectively.

Computer Numerical Control

Computer numerical control (CNC) is a method for automating control of machine tools through the use of software embedded in a microcomputer attached to the tool. It is commonly used in manufacturing for machining metal and plastic parts.

With CNC, each object to be manufactured gets a custom computer program, usually written in an international standard language called G-code, stored in and executed by the machine control unit (MCU), a microcomputer attached to the machine. The program contains the instructions and parameters the machine tool will follow, such as the feed rate of materials and the positioning and speed of the tool's components. Mills, lathes, routers, grinders and lasers are common machine tools whose operations can be automated with CNC. It can also be used to control non-machine tools, such as welding, electronic assembly and filament-winding machines.

Early in the process, engineers create a computer-aided design (CAD) drawing of the part to be manufactured, and then translate the drawing into G-code. The program is loaded onto the MCU and a human operator performs a test run without the raw material in place, to ensure proper positioning and performance. This step is important because incorrect speed or positioning can damage both the machine and the part.

CNC is considered to provide more precision, complexity and repeatability than is possible with manual machining. Other benefits include greater accuracy, speed and flexibility, as well as capabilities such as contour machining, which allows milling of contoured shapes, including those produced in 3D designs. On the other hand, CNC can be more expensive, require more maintenance than other production methods and compel companies to hire a skilled CNC programmer.

Some CNC systems are integrated with CAD and computer-aided manufacturing (CAM) software, which can speed the process of programming the MCU. Integration with ERP software and related applications, such as enterprise asset management software, can facilitate operational intelligence processes and help improve plant performance and maintenance.

CNC Machining Process

The term CNC stands for 'computer numerical control', and the CNC machining definition is that it is a subtractive manufacturing process which typically employs computerized controls and machine tools to remove layers of material from a stock piece—known as the blank or workpiece—and produces a custom-designed part. This process is suitable for a wide range of materials, including metals, plastics, wood, glass, foam, and composites, and finds application in a variety of industries, such as large CNC machining and CNC machining aerospace parts. When speaking in terms of the machine itself, the CNC machine definition is that it represents the actual programmable machine that is capable of autonomously performing the operations of CNC machining.

Subtractive manufacturing processes, such as CNC machining, are often presented in contrast to additive manufacturing processes, such as 3D printing, or formative manufacturing processes, such as liquid injection molding. While subtractive processes remove layers of material from the workpiece to produce custom shapes and designs, additive processes assemble layers of material to produce the desired form and formative processes deform and displace stock material into the desired shape. The automated nature of CNC machining enables the production of high precision and high accuracy, simple parts and the cost-effectiveness when fulfilling one-off and medium-volume production runs. However, while CNC machining demonstrates certain advantages over other manufacturing processes, the degree of complexity and intricacy attainable for part design and the cost-effectiveness of producing complex parts are limited.

Evolving from the numerical control (NC) machining process which utilized punched tape cards, CNC machining is a manufacturing process which utilizes computerized controls to operate and manipulate machine and cutting tools to shape stock material—e.g., metal, plastic, wood, foam, composite, etc.—into custom parts and designs. While the CNC machining process offers various capabilities and operations, the fundamental principles of the process remain largely the same throughout all of them. The basic CNC machining process includes the following stages:

- Designing the CAD model.
- Converting the CAD file to a CNC program.
- Preparing the CNC machine.
- Executing the machining operation.

CAD Model Design

The CNC machining process begins with the creation of a 2D vector or 3D solid part CAD design either in-house or by a CAD/CAM design Service Company. Computer-aided design (CAD) software allows designers and manufacturers to produce a model or rendering of their parts and products along with the necessary technical specifications, such as dimensions and geometries, for producing the part or product.

Designs for CNC machined parts are restricted by the capabilities (or inabilities) of the CNC machine and tooling. For example, most CNC machine tooling is cylindrical therefore the part geometries possible via the CNC machining process are limited as the tooling creates curved corner sections. Additionally, the properties of the material being machined, tooling design, and workholding capabilities of the machine further restrict the design possibilities, such as the minimum part thicknesses, maximum part size, and inclusion and complexity of internal cavities and features.

Once the CAD design is completed, the designer exports it to a CNC-compatible file format, such as STEP or IGES.

CAD File Conversion

The formatted CAD design file runs through a program, typically computer-aided manufacturing (CAM) software, to extract the part geometry and generates the digital programming code which will control the CNC machine and manipulate the tooling to produce the custom-designed part.

CNC machines used several programming languages, including G-code and M-code. The most well-known of the CNC programming languages, general or geometric code, referred to as G-code, controls when, where, and how the machine tools move—e.g., when to turn on or off, how fast to travel to a particular location, what paths to take, etc.—across the workpiece. Miscellaneous function code, referred to as M-code, controls the auxiliary functions of the machine, such as automating the removal and replacement of the machine cover at the start and end of production, respectively.

Once the CNC program is generated, the operator loads it to the CNC machine.

Machine Setup

Before the operator runs the CNC program, they must prepare the CNC machine for operation. These preparations include affixing the workpiece directly into the machine, onto machinery spindles, or into machine vises or similar workholding devices, and attaching the required tooling, such as drill bits and end mills, to the proper machine components.

Once the machine is fully set up, the operator can run the CNC program.

Machining Operation Execution

The CNC program acts as instructions for the CNC machine; it submits machine commands dictating the tooling's actions and movements to the machine's integrated computer, which operates and manipulates the machine tooling. Initiating the program prompts the CNC machine to begin the CNC machining process, and the program guides the machine throughout the process as it executes the necessary machine operations to produce a custom-designed part or product.

CNC machining processes can be performed in-house—if the company invests in obtaining and maintaining their own CNC equipment—or out-sourced to dedicated CNC machining service providers.

Types of CNC Machining Operations

CNC machining is a manufacturing process suitable for a wide variety of industries, including automotive, aerospace, construction, and agriculture, and able to produce a range of products, such as automobile frames, surgical equipment, airplane engines, and hand and garden tools. The process encompasses several different computer-controlled machining operations—including mechanical, chemical, electrical, and thermal processes—which remove the necessary material from the workpiece to produce a custom-designed part or product. This topic explores some of the most common mechanical CNC machining operations including:

- Drilling,
- Milling,
- Turning.

CNC Drilling

Drilling is a machining process which employs multi-point drill bits to produce cylindrical holes in the workpiece. In CNC drilling, typically the CNC machine feeds the rotating drill bit perpendicularly to the plane of the workpiece's surface, which produces vertically-aligned holes with diameters equal to the diameter of the drill bit employed for the drilling operation. However, angular drilling operations can also be performed through the use of specialized machine configurations and workholding devices. Operational capabilities of the drilling process include counter boring, countersinking, reaming, and tapping.

CNC Milling

Milling is a machining process which employs rotating multi-point cutting tools to remove material from the workpiece. In CNC milling, the CNC machine typically feeds the workpiece to the cutting tool in the same direction as the cutting tool's rotation, whereas in manual milling the machine feeds the workpiece in the opposite direction to the cutting tool's rotation. Operational capabilities of the milling process include face milling—cutting shallow, flat surfaces and flat-bottomed cavities into the workpiece—and peripheral milling—cutting deep cavities, such as slots and threads, into the workpiece.

CNC Turning

CNC Turning and Multi Spindle Machining.

Turning is a machining process which employs single-point cutting tools to remove material from the rotating workpiece. In CNC turning, the CNC machine—typically a lathe or turning machine—feeds the cutting tool in a linear motion along the surface of the rotating workpiece, removing material around the circumference until the desired diameter is achieved, to produce cylindrical parts with external and internal features, such as slots, tapers, and threads. Operational capabilities of the turning process include boring, facing, grooving, and thread cutting.

Table: Characteristics of Common CNC Machining Operations.

Machining Operation	Characteristics
Drilling	• Employs rotating multi-point drill bits. • Drill bit fed perpendicular or angularly to workpiece. • Produces cylindrical holes in workpiece.
Milling	• Employs rotating multi-point cutting tools. • Workpiece fed in same direction as cutting tool rotation. • Removes material from workpiece. • Produces broader range of shapes.
Turning	• Employs single-point cutting tools. • Rotates workpiece. • Cutting tool fed along the surface of the workpiece. • Removes material from the workpiece. • Produces round or cylindrical parts.

Other CNC Machining Operations

Other mechanical CNC machining operations include:

- Broaching,
- Sawing,
- Grinding,
- Honing,
- Lapping.

Importance of CNC in Manufacturing

Computer numerical control is an extremely high-tech and important process in today's world of manufacturing. It takes skilled engineers to program the files used by the microprocessor within a machine tool to make a machine function properly and efficiently. It's amazing how quickly various goods can be produced, but CNC technology is an important backbone that makes all the difference in these processes.

Motion Control Systems

A motion control system is any system that controls energy flow to generate mechanical motion with some desired properties. Motion control is a bit of a specialty in automated control systems, and its use is not basic, as it can provide advanced machine functionality. It provides the means to move the machine tooling or the part itself in a controlled, and often precision, rotary or linear manner.

Historically, motion control had its own controller and application software with motion-control algorithms to get the work done. Today, much of the motion control can be performed in the main control-system controller, if chosen carefully, but motion still requires specialty amplifiers/drives, motors, position feedback devices and precision mechanical linkage or actuators.

A motion controller contains the motion profiles and target positions for the application and creates the trajectories for the motor and actuator. Motion control is often closed loop, so it monitors the actual path and corrects for position or velocity errors. The motion controller often outputs position commands via +/-10 Vdc signals for a servo motor drive or step and direction signals to a stepper motor drive. Industrial Ethernet or similar protocols are used to command many modern drives.

A variety of motion controllers are available. Categories include PLCs, PACs, PC-based, stand-alone and microcontrollers. Modern, programmable logic controllers (PLCs) and programmable automation controllers (PACs) have a limited to wide range of motion-control capabilities. PC-based systems can expand on many of the PLC and PAC capabilities. Stand-alone controllers can be part of the drive or separate with a focus on motion control. Individual microcontrollers can be attached to a part of a small drive or can be directly integrated into an all-in-one drive-and-motor combination.

Regardless of the category of microcontroller selected, some key specifications to consider are the type of motion supported, number of axes, scan rate and feedback resolution. These play a big role in the controller and other motion-control products selected.

Some automation controllers only provide limited motion control, such as a couple of high-speed inputs for feedback signals and a few high-speed outputs for pulse and direction commands to stepper motor drives. The controller may also just control a motor's rpm via a Modbus communication connection to a variable frequency drive (VFD) or turn on a discrete output that runs a motion profile stored in stand-alone drive, and there are many other options.

The drives, including variable frequency, stepper and servo, receive the low-level analog, discrete or communication signals from the selected controller and amplify them to provide high-power voltage and current to the motor. The drives are closely matched to the motor specified, but there are many types including variable-frequency drives, stepper-motor drives, analog-servo drives, digital-servo drives, brushed or brushless dc drives and others.

The controllers are telling the drives where to move the motors that turn the electrical energy they receive into mechanical energy. The motors produce the torque to create the rotational force or linear force when connected to an actuator to rotate at a setpoint speed or move to a commanded position. There are many types of dc and ac motors.

DC motors include both brush and brushless types. The brushed type, such as permanent magnet, series wound, shunt wound and compound wound, use internal commutation and are connected directly to a dc power source to operate. Brushless dc motors, such as stepper motors, need an electronic drive to operate.

The dc motor has many desirable characteristics; a notable one is that it has maximum torque at zero speed. The dc motor is typically simple to install with a wide speed control range, high starting torque and quick acceleration, stopping and reversing.

AC motors include both induction and synchronous types. The induction motor, often used to run pumps, fans and conveyors, can be run directly from line voltage through a motor starter, using a variable-frequency drive or a soft start. The synchronous motor's rotation is closely synchronized with the supply current frequency.

AC motors offer low power demand when starting, and the acceleration, speed, starting current and torque limit can be controlled.

A servo motor can be many types of motors. Many are dc motors, and brushed dc versions are the simplest of servo motors. However, many of the servo motors used in automation are brushless dc motors. The ac brushless servo motor, with permanent magnets, is another option that provides higher performance.

There are many feedback sensors used in motion control. At its simplest, tachometers provide velocity feedback. Photometers provide an analog position feedback, and quadrature encoders provide a digital position relative to a starting, or home, position. Both absolute encoders and resolvers provide absolute position measurement, even after cycling power off and on.

There are many mechanical components to transform the rotary motion into the desired motion, often linear. The bearings reduce the friction and support radial and axial loads. Gearboxes multiply the torque and help to match load inertia while reducing the speed of the output shaft. Ball screws and lead screws connect to the output shaft to convert the rotational motion to linear motion. And linkages tie it all together from the output shaft to actuator, chain, sprocket, belt and pulley, depending on the application.

High Performance Motion Control

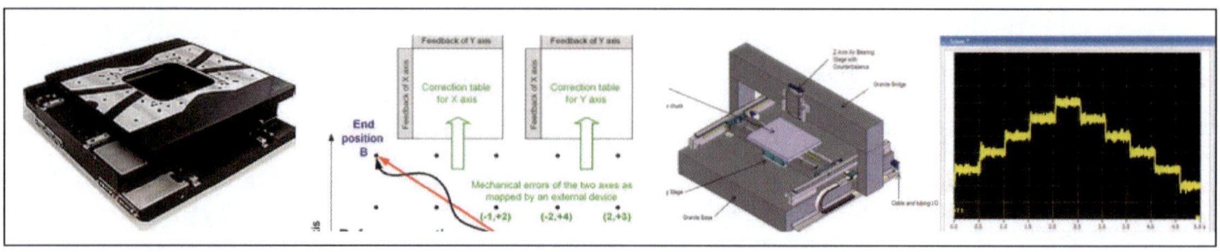

High performance motion control plays a key part in positioning and handling applications, such as those in semiconductor manufacturing, laser and materials processing, optical inspection systems, additive manufacturing or industrial digital printing. In addition to providing reliability and robustness, deterministic behavior with minimized latency, high-speed synchronization with external devices, such as lasers and cameras, is required to achieve sub-micrometer path accuracy, exact positioning repeatability, high dynamics, and throughput. Safety, ease of operation, and state-of-the-art connectivity for integration into industrial network environments are further requirements.

The higher the dynamics and throughput targets of a multi-axis precision motion system, the more holistic the design approach that has to be taken to achieve success. The best positioning mechanics cannot follow multi-dimensional motion profiles without an industry grade motion controller that will synchronize the individual axes on a sub-millisecond time scale, while at the same time communicating with a higher-level automation environment. A systems approach where mechanics, control electronics, and software are designed by the same team, provides obvious advantages for the user.

Quality Control Systems

Quality Control (QC) may be defined as a system that is used to maintain a desired level of quality in a product or service. It is a systematic control of various factors that affect the quality of the product. It depends on materials, tools, machines, type of labor, working conditions etc. QC is a broad term, it involves inspection at particular stage but mere inspection does not mean QC. As opposed to inspection, in quality control activity emphasis is placed on the quality future production. Quality control aims at prevention of defects at the source, relies on effective feedback system and corrective action procedure. Quality control uses inspection as a valuable tool.

According to Juran "Quality control is the regulatory process through which we measure actual quality performance, compare it with standards, and act on the difference". Another definition of quality control is from ANSI/ASQC standard quality control is defined as "The operational techniques and the activities which sustain a quality of product or service that will satisfy given needs; also the use of such techniques and activities".

Alford and Beatty define QC as "In the broad sense, quality control is the mechanism by which products are made to measure up to specifications determined from customers, demands and transformed into sales engineering and manufacturing requirements, it is concerned with making things right rather than discovering and rejecting those made wrong".

Types of Quality Control

QC is not a function of any single department or a person. It is the primary responsibility of any supervisor to turn out work of acceptable quality. Quality control can be divided into three main sub-areas, those are:

- Off-line quality control: Its procedure deal with measures to select and choose controllable product and process parameters in such a way that the deviation between the product or process output and the standard will be minimized. Much of this task is accomplished through product and process design. Example: Taguchi method, principles of experimental design etc.
- Statistical process control: SPC involves comparing the output of a process or a service with a standard and taking remedial actions in case of a discrepancy between the two. It also involves determining whether a process can produce a product that meets desired specification or requirements. On-line SPC means that information is gathered about the product, process, or service while it is functional. The corrective action is taken in that operational phase. This is real-time basis.
- Acceptance sampling plans: A plan that determines the number of items to sample and the acceptance criteria of the lot, based on meeting certain stipulated

conditions (such as the risk of rejecting a good lot or accepting a bad lot) is known as an acceptance sampling plan.

Steps in Quality Control

Following are the steps in quality control process:

- Formulate quality policy.
- Set the standards or specifications on the basis of customer's preference, cost and profit.
- Select inspection plan and set up procedure for checking.
- Detect deviations from set standards of specifications.
- Take corrective actions or necessary changes to achieve standards.
- Decide on salvage method i.e., to decide how the defective parts are disposed of, entire scrap or rework.
- Coordination of quality problems.
- Developing quality consciousness both within and outside the organization.
- Developing procedures for good vendor-vendee relations.

Objectives of Quality Control

Following are the objectives of quality control:

- To improve the companies income by making the production more acceptable to the customers, i.e., by providing long life, greater usefulness, maintainability etc.
- To reduce companies cost through reduction of losses due to defects.
- To achieve interchangeability of manufacture in large scale production.
- To produce optimal quality at reduced price.
- To ensure satisfaction of customers with productions or services or high quality level, to build customer goodwill, confidence and reputation of manufacturer.
- To make inspection prompt to ensure quality control.
- To check the variation during manufacturing.

The broad areas of application of quality control are incoming material control, process control and product control.

Benefits of Quality Control

- Improving the quality of products and services.
- Increasing the productivity of manufacturing processes, commercial business, and corporations.
- Reducing manufacturing and corporate costs.
- Determining and improving the marketability of products and services.
- Reducing consumer prices of products and services.
- Improving and assuring on time deliveries and availability.
- Assisting in the management of an enterprise.

Seven Tools for Quality Control

To make rational decisions using data obtained on the product, or process, or from the consumer, organizations use certain graphical tools. These methods help us learn about the characteristics of a process, its operating state of affairs and the kind of output we may expect from it. Graphical methods are easy to understand and provide comprehensive information; they are a viable tool for the analysis of product and process data. These tools are effect on quality improvement. The seven quality control tools are:

- Pareto charts,
- Check sheets,
- Cause and effect diagram,
- Scatter diagrams,
- Histogram,
- Graphs or flow charts,
- Control charts.

Pareto Charts

Pareto charts help prioritize by arranging them in decreasing order of importance. In an environment of limited resources these diagrams help companies to decide on the order in which they should address problems. The Pareto analysis can be used to identify the problem in a number of forms:

- Analysis of losses by material (number or past number).
- Analysis of losses by process i.e., classification of defects or lot rejections in terms of the process.

- Analysis of losses by product family.
- Analysis by supplier across the entire spectrum of purchases.
- Analysis by cost of the parts.
- Analysis by failure mode.

Example: A Pareto chart of reasons for poor quality. Poor design will be the major reason, as indicated by 64%. Thus, this is the problem that the manufacturing unit should address first.

- Poor Design,
- Defective Parts,
- Operator Error,
- Wrong Dimensions,
- Surface Abrasion,
- Machine Calibrations,
- Defective Material.

Check Sheets

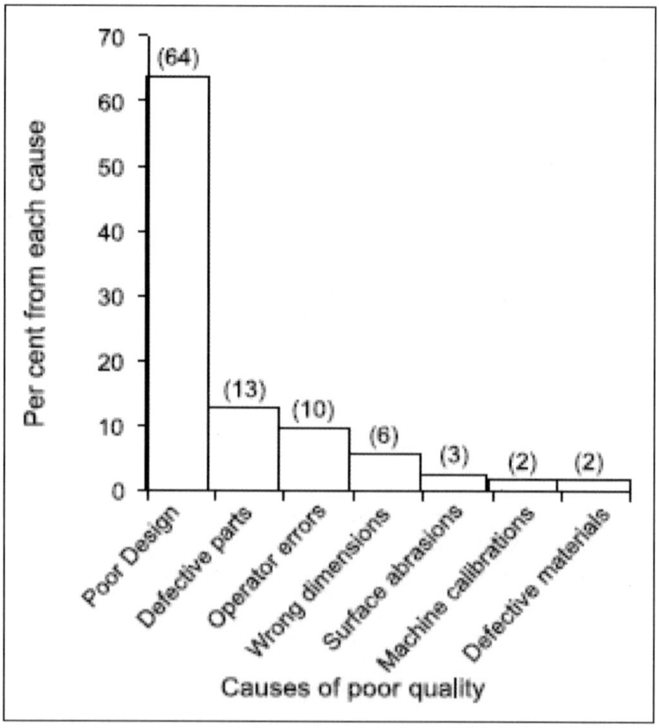

Pareto chart.

Check sheets facilitate systematic record keeping or data collection observations are recorded as they happen which reveals patterns or trends. Data collection through the use of a checklist is often the first step in analysis of quality problem. A checklist is a form used to record the frequency of occurrence of certain product or service characteristics related to quality. The characteristics may be measurable on a continuous scale such as weight, diameter, time or length.

Example: The table is a check sheet for an organization's computer related problems.

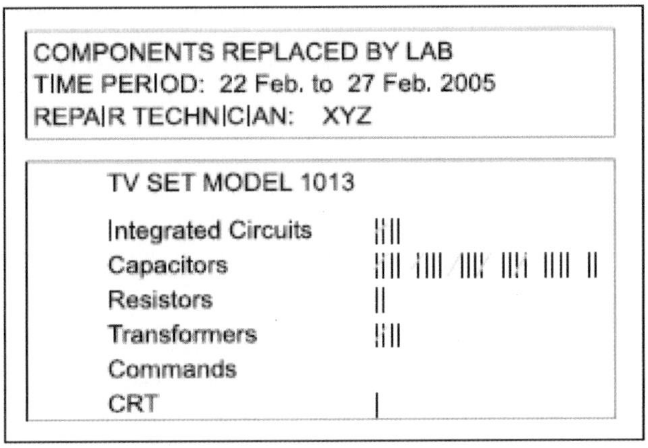

Checklist.

Cause and Effect Diagram

It is sometimes called as Fish-bone diagram. It is first developed by Kaorv Ishikawa in 1943 and is sometimes called as Ishikawa diagram. The diameter helps the management trace customer complaints directly to the operations involved. The main quality problem is referred to Fish-head; the major categories of potential cause structural bones and the likely specific causes to ribs. It explores possible causes of problems, with the intention being to discover the root causes. This diagram helps identify possible reasons for a process to go out of control as well as possible effects on the process.

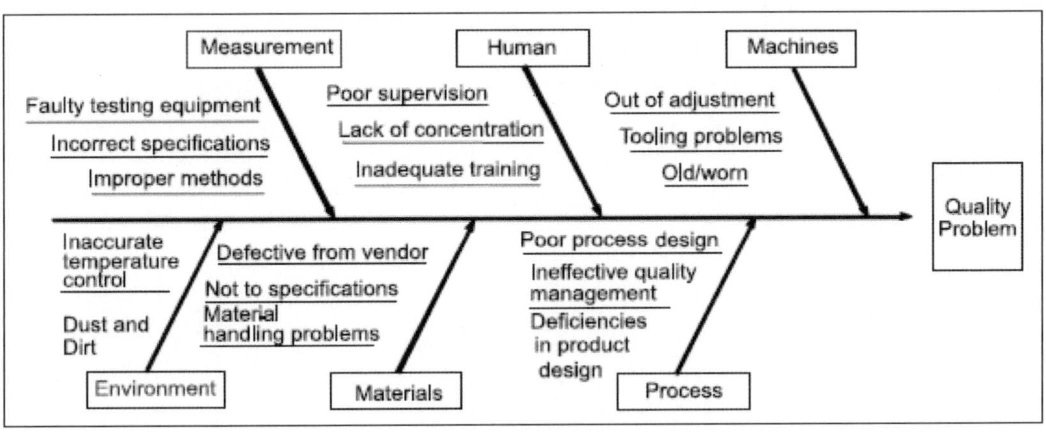

Fishbone diagram.

Scatter Diagram

It often indicates the relationship between two variables. They are often used as follow-ups to a cause and effect analysis to determine whether a stated cause truly does impact the quality characteristics.

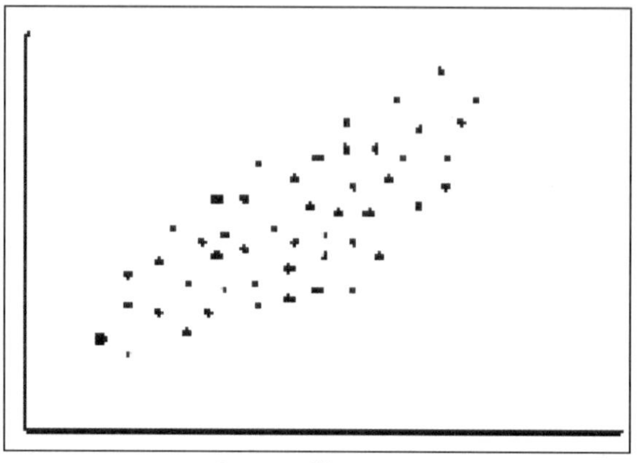

Scatter diagram.

Example: The plots advertising expenditure against company sales and indicates a strong positive relationship between the two variables. As the level of advertising expenditure increases sales tend to increase.

Histogram Bar Charts

It displays the large amounts of data that are difficult to interpret in their raw form. A histogram summarizes data measured on a continuous scale showing the frequency distribution of some quality characteristics (in statistical terms the central tendency and the dispersion of the data).

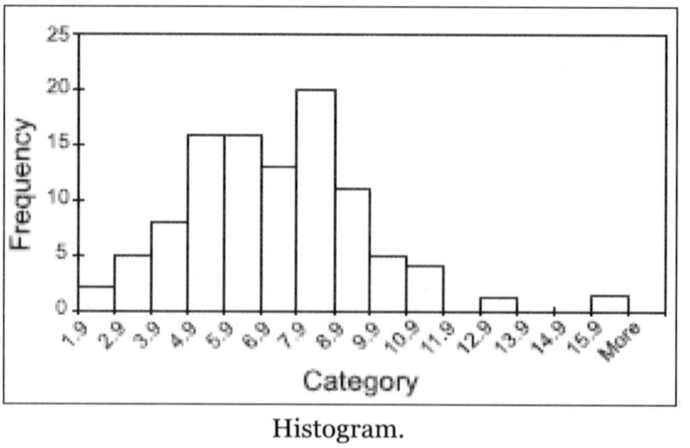

Histogram.

Often the mean of the data is indicated on the histogram. A bar chart is a series of bare representing the frequency of occurrence of data characteristics, the bar height indicates the number of times a particular quality characteristic was observed.

Flow Charts Graphs

It shows the sequence of events in a process. They are used for manufacturing and service operations. Flow charts are often used to diagram operational procedures to simplify the system. They can identify bottlenecks, redundant steps and non-value added activities. A realistic flow chart can be constructed by using the knowledge of the person who are directly involved in the particular process. The flow chart can be identifies where delays can occur.

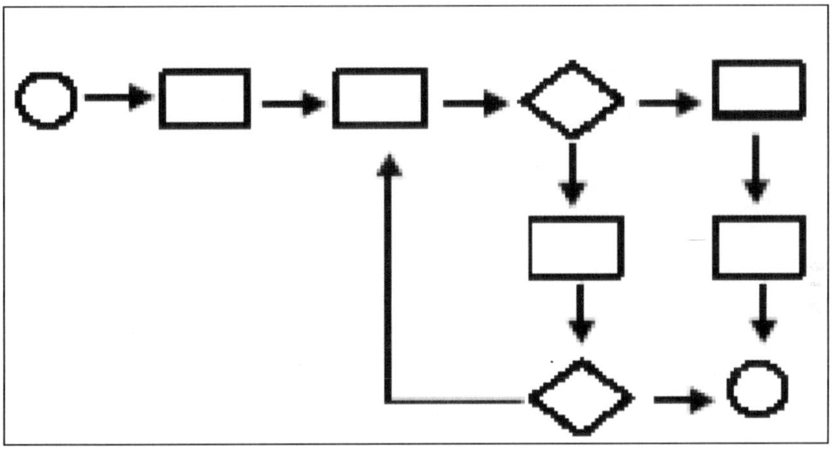

Flowchart.

Control Charts

It distinguishes special causes of variations from common causes of variation. They are used to monitor and control process on an ongoing basis. A typical control chart plots a selected quality characteristic found from sub-group of observations as a function of sample number.

Characteristics such as sample average, sample range and sample proportion of non-conforming units are plotted. The centre line on a control chart represents the average value of characteristics being plotted. Two limits known as the upper control limit (UCL) and lower control limit (LCL) are also shown on control charts. These limits are constructed so that if the process is operating under a stable system of chance causes, the problem of an observation falling outside these limits is quite small. The following figure shows a generalized representation of a control chart.

Control chart shows the performance of a process from two points of view. First, they show a snapshot of the process at the moment the data are collected.

Second, they show the process trend as time progresses. Process trends are important because they help in identifying the out-of-control status if it actually exists. Also, they help to detect variations outside the normal operational limits, and to identify the cause of variations. Figure shows a generalized representation of a control chart.

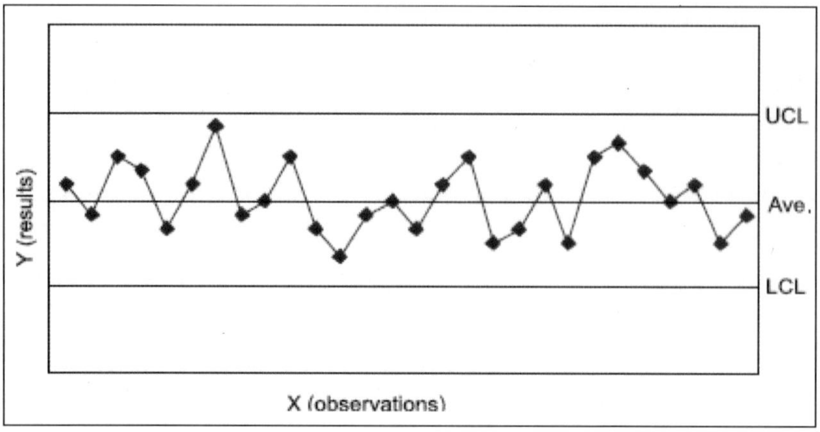

Control charts.

Causes of Variation in Quality

The variation in the quality of product in any manufacturing process is broadly classified as:

- Chance causes,
- Assignable causes.

Chance Causes

The chance causes are those causes which are inherit in manufacturing process by virtue of operational and constructional features of the equipment's involved in a manufacturing process.

This is because of:

- Machine vibrations,
- Voltage variations,
- Composition variation of material, etc.

They are difficult to trace and difficult to control, even under best condition of production. Even though, it is possible to trace out, it is not economical to eliminate. The chance causes results in only a minute amount of variation in process. Variation in chance causes is due to internal factors only the general pattern of variation under chance causes will follow a stable statistical distribution (normal distribution). Variation within the control limits means only random causes are present.

Assignable Causes

These are the causes which creates ordinary variation in the production quality.

Assignable cause's variation can always be traced to a specific quality. They occur due to:

- Lack of skill in operation.
- Wrong maintenance practice.
- New vendors.
- Error in setting jigs and fixtures.
- Raw material defects.

Variation due to these causes can be controlled before the defective items are produced. Any one assignable cause can result in a large amount of variation in process. If the assignable causes are present, the system will not follow a stable statistical distribution. When the actual variation exceeds the control limits, it is a signal that assignable causes extend the process and process should be investigated.

Proportional–Integral–Derivative Controller

Proportional-Integral-Derivative (PID) control is the most common control algorithm used in industry and has been universally accepted in industrial control. The popularity of PID controllers can be attributed partly to their robust performance in a wide range of operating conditions and partly to their functional simplicity, which allows engineers to operate them in a simple, straightforward manner.

As the name suggests, PID algorithm consists of three basic coefficients; proportional, integral and derivative which are varied to get optimal response. Closed loop systems, the theory of classical PID and the effects of tuning a closed loop control system are discussed in this paper.

Control System

The basic idea behind a PID controller is to read a sensor, then compute the desired actuator output by calculating proportional, integral, and derivative responses and summing those three components to compute the output. Before we start to define the parameters of a PID controller, we shall see what a closed loop system is and some of the terminologies associated with it.

Closed Loop System

In a typical control system, the process variable is the system parameter that needs to be controlled, such as temperature (°C), pressure (psi), or flow rate (liters/minute). A sensor is used to measure the process variable and provide feedback to the control system. The set point is the desired or command value for the process variable, such as 100 degrees Celsius in the case of a temperature control system. At any given moment, the difference between the process variable and the set point is used by the control system algorithm (compensator), to determine the desired actuator output to drive the system (plant). For instance, if the measured temperature process variable is 100 °C and the desired temperature set point is 120 °C, then the actuator output specified by the control algorithm might be to drive a heater. Driving an actuator to turn on a heater causes the system to become warmer, and results in an increase in the temperature process variable. This is called a closed loop control system, because the process of reading sensors to provide constant feedback and calculating the desired actuator output is repeated continuously and at a fixed loop rate as illustrated in figure.

In many cases, the actuator output is not the only signal that has an effect on the system. For instance, in a temperature chamber there might be a source of cool air that sometimes blows into the chamber and disturbs the temperature. Such a term is referred to as disturbance. We usually try to design the control system to minimize the effect of disturbances on the process variable.

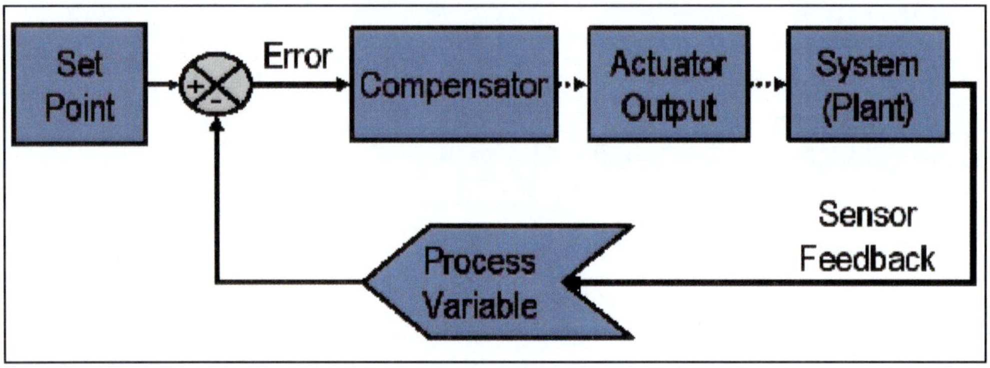

Block diagram of a typical closed loop system.

Terminologies

The control design process begins by defining the performance requirements. Control system performance is often measured by applying a step function as the set point command variable, and then measuring the response of the process variable. Commonly, the response is quantified by measuring defined waveform characteristics. Rise Time is the amount of time the system takes to go from 10% to 90% of the steady-state, or final, value. Percent Overshoot is the amount that the process variable overshoots the final value, expressed as a percentage of the final value. Settling time is the time required for the process variable to settle to within a certain percentage (commonly 5%) of the final value. Steady-State Error is the final difference between the process variable and set point.

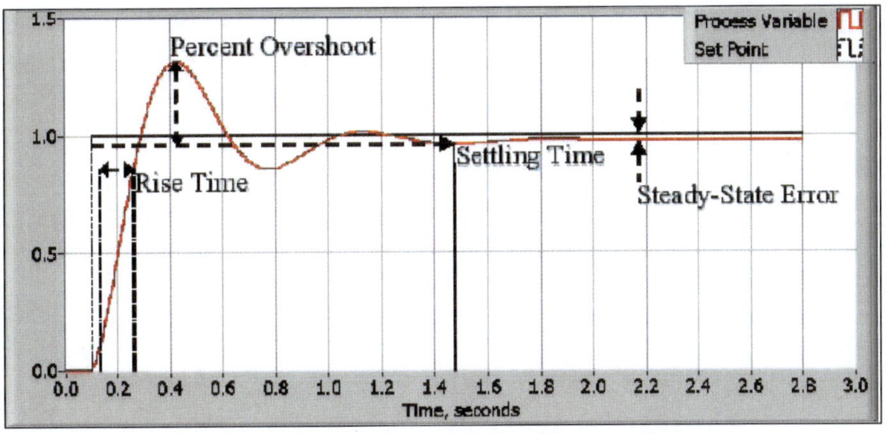

Response of a typical PID closed loop system.

After using one or all of these quantities to define the performance requirements for a control system, it is useful to define the worst case conditions in which the control system will be expected to meet these design requirements. Often times, there is a disturbance in the system that affects the process variable or the measurement of the process variable. It is important to design a control system that performs satisfactorily during worst case conditions. The measure of how well the control system is able to overcome the effects of disturbances is referred to as the disturbance rejection of the control system.

In some cases, the response of the system to a given control output may change over time or in relation to some variable. A nonlinear system is a system in which the control parameters that produce a desired response at one operating point might not produce a satisfactory response at another operating point. For instance, a chamber partially filled with fluid will exhibit a much faster response to heater output when nearly empty than it will when nearly full of fluid. The measure of how well the control system will tolerate disturbances and nonlinearities is referred to as the robustness of the control system.

Some systems exhibit an undesirable behavior called deadtime. Deadtime is a delay between when a process variable changes, and when that change can be observed. For instance, if a temperature sensor is placed far away from a cold water fluid inlet valve, it will not measure a change in temperature immediately if the valve is opened or closed. Deadtime can also be caused by a system or output actuator that is slow to respond to the control command, for instance, a valve that is slow to open or close. A common source of deadtime in chemical plants is the delay caused by the flow of fluid through pipes.

Loop cycle is also an important parameter of a closed loop system. The interval of time between calls to a control algorithm is the loop cycle time. Systems that change quickly or have complex behavior require faster control loop rates.

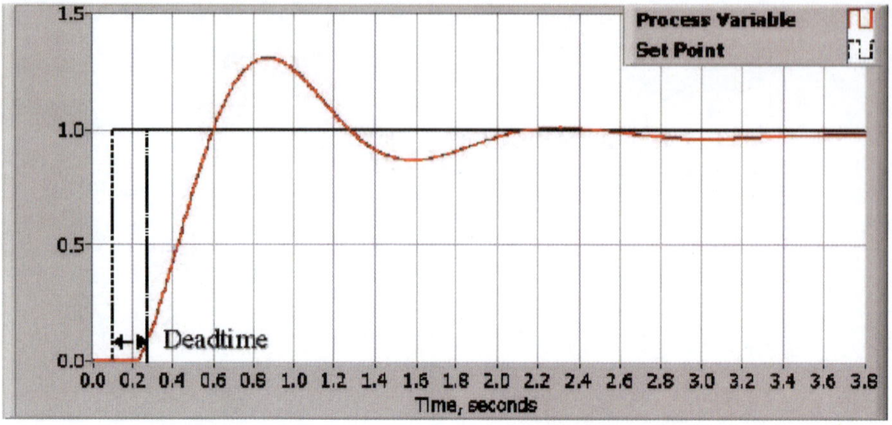

Response of a closed loop system with deadtime.

Once the performance requirements have been specified, it is time to examine the system and select an appropriate control scheme. In the vast majority of applications, a PID control will provide the required results.

PID Theory

Proportional Response

The proportional component depends only on the difference between the set point and the process variable. This difference is referred to as the Error term. The proportional gain (K_c) determines the ratio of output response to the error signal. For instance, if the error term has a magnitude of 10, a proportional gain of 5 would produce a proportional response of 50. In general, increasing the proportional gain will increase the speed of the control system response. However, if the proportional gain is too large, the process variable will begin to oscillate. If K_c is increased further, the oscillations will become larger and the system will become unstable and may even oscillate out of control.

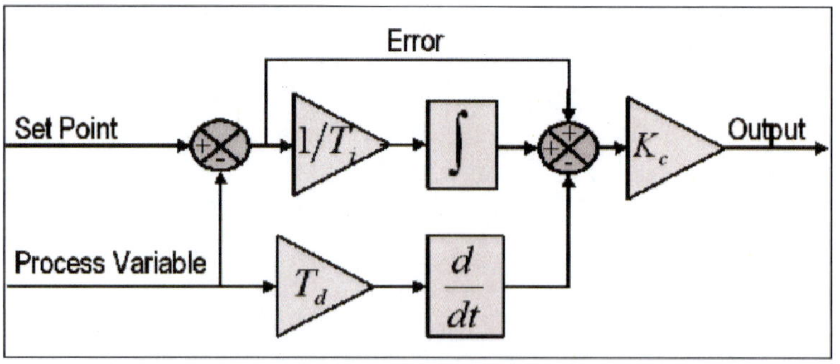

Block diagram of a basic PID control algorithm.

Integral Response

The integral component sums the error term over time. The result is that even a small error term will cause the integral component to increase slowly. The integral response will continually increase over time unless the error is zero, so the effect is to drive the Steady-State error to zero. Steady-State error is the final difference between the process variable and set point. A phenomenon called integral windup results when integral action saturates a controller without the controller driving the error signal toward zero.

Derivative Response

The derivative component causes the output to decrease if the process variable is increasing rapidly. The derivative response is proportional to the rate of change of the process variable. Increasing the derivative time (Td) parameter will cause the control system to react more strongly to changes in the error term and will increase the speed of the overall control system response. Most practical control systems use very small derivative time (Td), because the Derivative Response is highly sensitive to noise in the process variable signal. If the sensor feedback signal is noisy or if the control loop rate is too slow, the derivative response can make the control system unstable.

PID Modes of Control

Proportional (P) Response

Term ‘P’ is proportional to the actual value of the error. If the error is large, control output is also large and if the error is small control output is also small, but gain factor (K_p) is

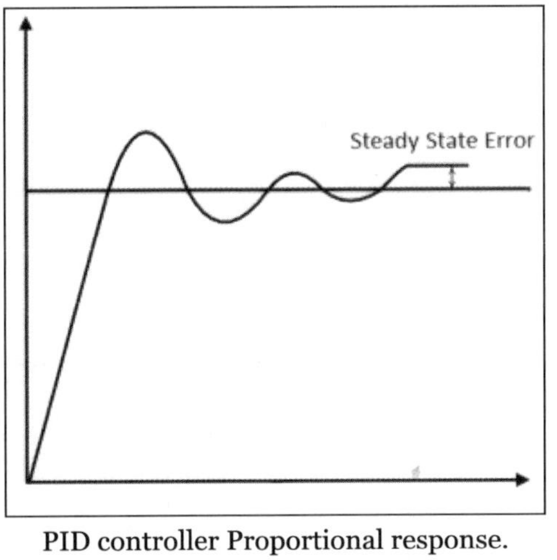

PID controller Proportional response.

Also taking in to account. Speed of response is also directly proportional to proportional gain factor (K_p). So, the speed of response is increased by increasing the value of K_p but if K_p is increased beyond normal range, process variable starts oscillating at high rate and make system unstable.

$y(t) \propto e(t)$

$y(t) = ki * e(t)$

Where K_p is proportional gain factor.

Here, the resulting error is multiplied with proportionality gain factor (proportional constant) as shown in above equation. If only P controller is used, at that time, it requires manual reset because it maintains steady state error (offset).

Integral (I) Response

Integral controller is generally used to decrease the steady state error. Term 'I' is integrate (with respect to time) to the actual value of the error. Because of integration, very small value of error, results very high integral response. Integral controller action continues to change until error becomes zero.

$y(t) \propto \int e(t)$

$y(t) = ki \int e(t)$

Where K_i is a proportional gain factor.

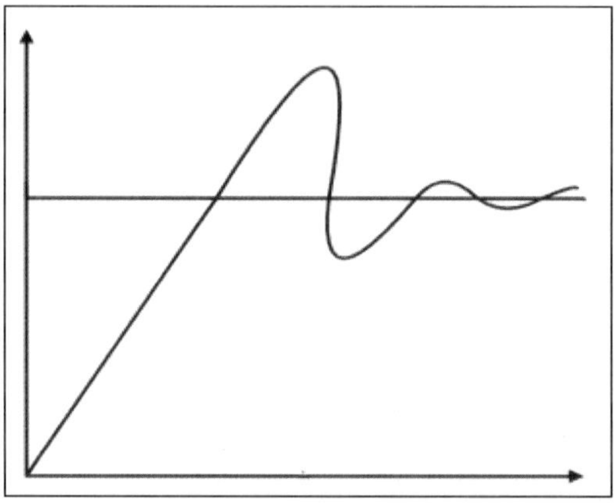

PID controller Integral response.

Integral gain is inversely proportional to the speed of response, increasing k_i, decrease the speed of response. Proportional and Integral controllers are used combined (PI controller) for good speed of response and steady state response.

Derivative (D) Response

Derivative controller is used to with combination of PD or PID. It never used alone, because if error is constant (non-zero), output of the controller will be zero. In this situation, controller behave life zero error, but in actual there are some error (constant). Output of derivative controller is directly proportional to the rate of change of error with respect to time as shown in equation. By removing sign of proportionality, we get derivative gain constant (k_d). Generally, Derivative controller is used when processor variables starts oscillating or changes at a very high rate of speed. D-controller is also used to anticipate the future behaviour of the error by error curve. Mathematical equation is as shown below:

$$y(t) \propto de(t)/dt$$

$$y(t) = Kd * de(t)/dt$$

Where K_d is a proportional gain factor.

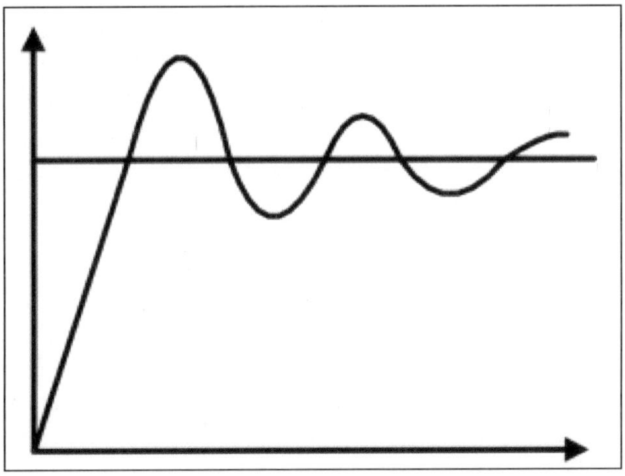

PID controller Derivative response.

Proportional and Integral controller

This is a combination of P and I controller. Output of the controller is summation of both (proportional and integral) responses. Mathematical equation is as shown in below:

$$y(t) \propto (e(t) + \int e(t)\, dt)$$

$$y(t) = kp * e(t) + ki \int e(t)\, dt$$

Proportional and Derivative Controller

This is a combination of P and D controller. Output of controller is summation of proportional and derivative responses. Mathematical equation of PD controller is as shown below:

$$y(t) \propto (e(t) + de(t)/dt)$$

$$y(t) = kp * e(t) + kd * de(t)/dt$$

Proportional, Integral and Derivative Controller

This is a combination of P, I and D controller. Output of controller is summation of proportional, integral and derivative responses. Mathematical equation of PD controller is as shown below:

$$y(t) \propto (e(t) + \int e(t)\,dt + de(t)/dt)$$

$$y(t) = kp * e(t) + ki \int e(t)\,dt + kd * de(t)/dt$$

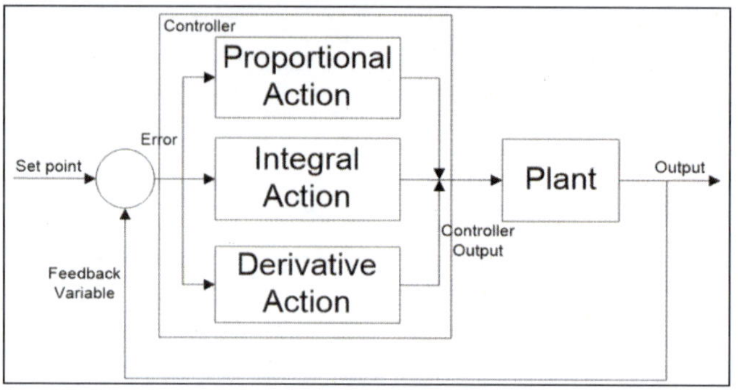

PID controller Block diagram.

Thus, by combining this proportional, integral and derivative control response, form a PID controller.

Working of PID Controller

With the use of low cost simple ON-OFF controller only two control states are possible, like fully ON or fully OFF. It is used for limited control application where these two control states are enough for control objective. However oscillating nature of this control limits its usage and hence it is being replaced by PID controllers.

PID controller maintains the output such that there is zero error between process variable and set point/ desired output by closed loop operations. PID uses three basic control behaviors that are explained below.

P-controller

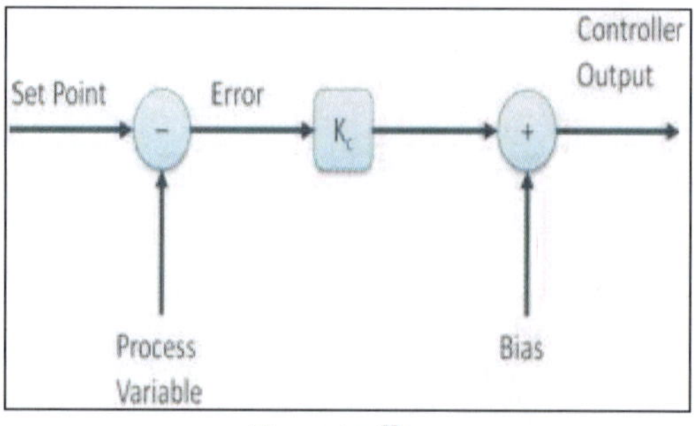

P-controller.

Proportional or P- controller gives output which is proportional to current error e (t). It compares desired or set point with actual value or feedback process value. The resulting error is multiplied with proportional constant to get the output. If the error value is zero, then this controller output is zero.

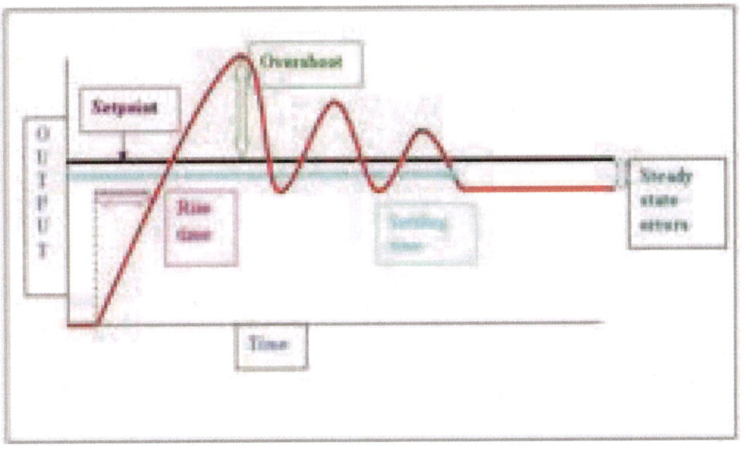

P-Controller Response.

This controller requires biasing or manual reset when used alone. This is because it never reaches the steady state condition. It provides stable operation but always maintains the steady state error. Speed of the response is increased when the proportional constant K_c increases.

I-controller

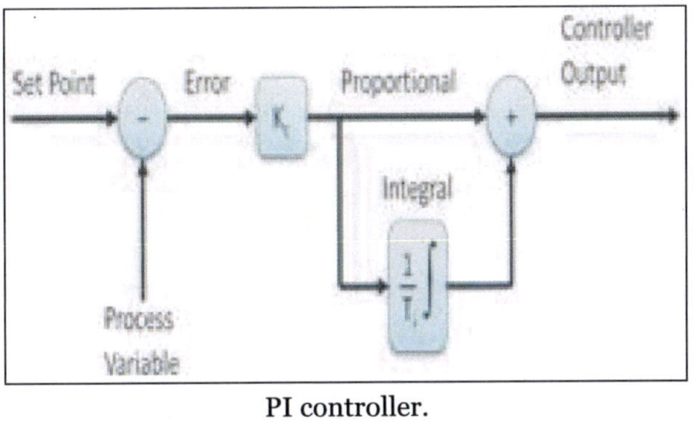

PI controller.

Due to limitation of p-controller where there always exists an offset between the process variable and set point, I-controller is needed, which provides necessary action to eliminate the steady state error. It integrates the error over a period of time until error value reaches to zero. It holds the value to final control device at which error becomes zero.

Integral control decreases its output when negative error takes place. It limits the speed of response and affects stability of the system. Speed of the response is increased by decreasing integral gain Ki.

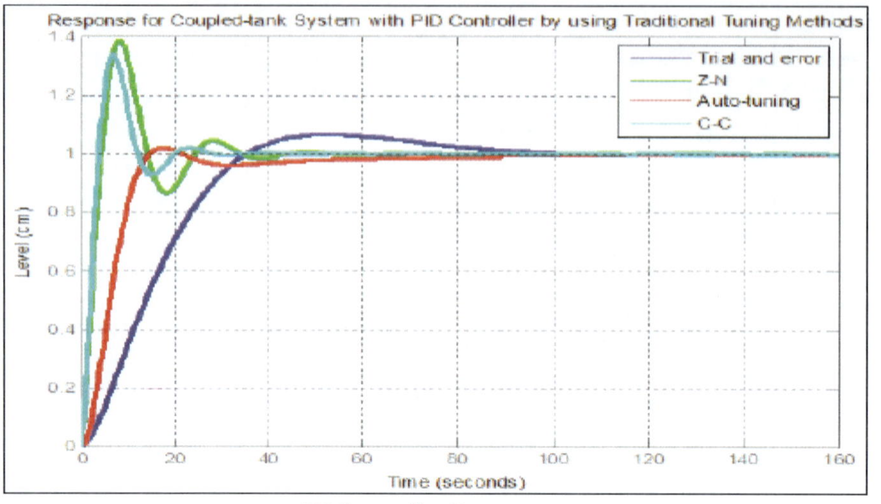

PI Controller Response.

In above figure, as the gain of the I-controller decreases, steady state error also goes on decreasing. For most of the cases, PI controller is used particularly where high speed response is not required.

While using the PI controller, I-controller output is limited to somewhat range to overcome the integral wind up conditions where integral output goes on increasing even at zero error state, due to nonlinearities in the plant.

D-controller

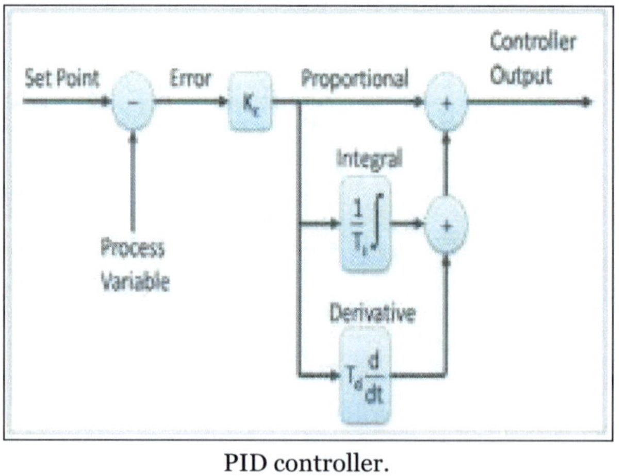

PID controller.

I-controller doesn't have the capability to predict the future behavior of error. So it reacts normally once the set point is changed. D-controller overcomes this problem by anticipating future behavior of the error. Its output depends on rate of change of error with respect to time, multiplied by derivative constant. It gives the kick start for the output thereby increasing system response.

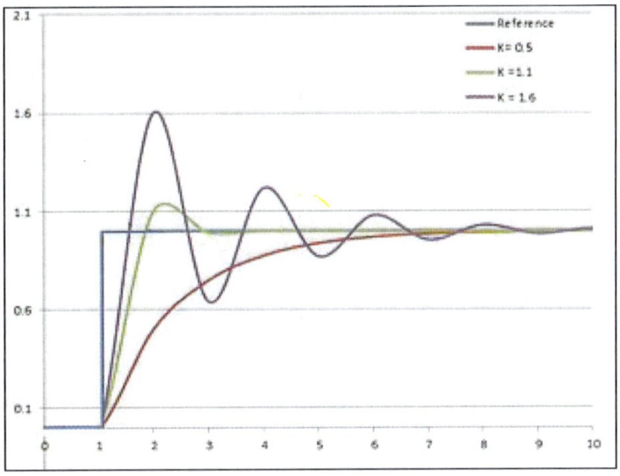

PID Controller Response.

In the above figure response of D controller is more, compared to PI controller and also settling time of output is decreased. It improves the stability of system by compensating phase lag caused by I-controller. Increasing the derivative gain increases speed of response.

Tuning Methods for PID Controller

For desired output, this controller must be properly tuned. The process of getting ideal response from the PID controller by PID setting is called tuning of controller. PID setting means set the optimal value of gain of proportional (k_p), derivative (k_d) and integral (k_i) response. PID controller is tuned for disturbance rejection means staying at a given setpoint and command tracking, means if setpoint is change, output of controller will follow new setpoint. If controller is properly tuned, output of controller will follow variable setpoint, with less oscillation and less damping.

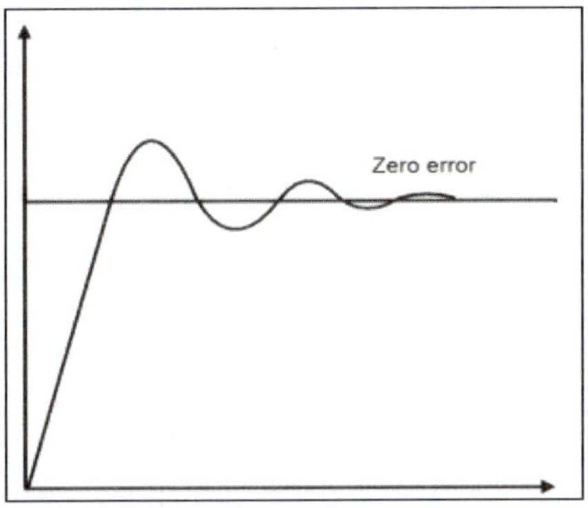

Disturbance Rejection by PID controller tunner.

There are several methods for tuning PID controller and getting desired response. Methods for tuning controller is as below:

- Trial and error method: Trial and error method is also known as manual tuning method and this method is simplest method. In this method, first increase the value of kp until system reaches to oscillating response but system should not make unstable and keep value of kd and ki zero. After that, set value of ki in such a way that, oscillation of system is stops. After that set the value of kd for fast response.

- Process reaction curve technique: This method is also known as Cohen-Coon tuning method. In this method first generate a process reaction curve in response to a disturbance. By this curve we can calculate the value of controller gain, integral time and derivative time. This curve is identified by performing manually in open loop step test of the process. Model parameter can find by initial step percentage disturbance. From this curve we have to find slop, dead time and rise time of curve which is nothing but the value of kp, ki and kd.

- Zeigler-Nichols method: In this method also first set the value of ki and kd zero. The proportional gain (kp) is increase until it reaches at the ultimate gain (ku). ultimate gain is nothing but it is a gain at which output of loop starts to oscillates. This ku and the oscillation period Tu are used to derive gain of PID controller from below table.

Type of Controller	k_p	k_i	k_d
P	$0.5\ k_u$	-	-
PI	$0.45\ k_u$	$0.54\ k_u/T_u$	-
PID	$0.60\ k_u$	$1.2\ k_u/T_u$	$3\ k_u T_u/40$

- Relay method: This method is also known as Astrom-Hugglund method. Here output is switched between two values of the control variable but these values are chosen in such a way that process must cross the setpoint. When process variable is less than setpoint, the control output is set to the higher value. When process value is greater than setpoint, the control output is set to the lower value and output waveform is formed. The period and amplitude of this oscillatory waveform is measured and used to determine ultimate gain ku and period Tu which is used in above method.

- Using software: For PID tuning and loop optimization, software packages are available. These software packages collect data and make a mathematical model of system. By this model, software finds an optimal tuning parameter from reference changes.

Structure of PID Controller

PID controllers are designed based on the microprocessor technology. Different manufactures use different PID structure and equation. Most common used PID equations are; parallel, ideal and series PID equation.

In parallel PID equation, proportional, integral and derivative actions are working separately with each other and combine effect of these three actions are act in the system. Block diagram of this type of PID is as shown below:

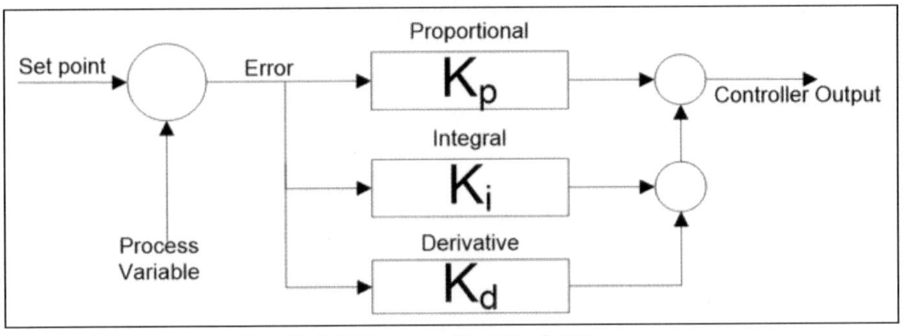

Parallel PID controller Block Diagram.

In ideal PID equation, gain constant k_p is distributed to all term. So, changes in k_p affects all other terms in the equation.

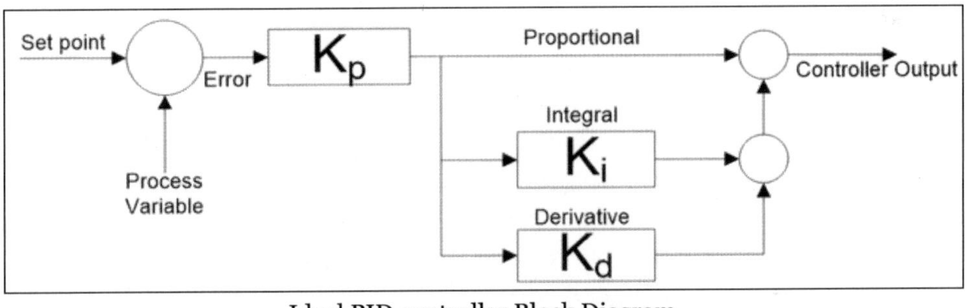

Ideal PID controller Block Diagram.

In series PID equation, gain constant k_p is distributed to all terms same as ideal PID equation, but in this equation integral and derivative constant have an effect on proportional action.

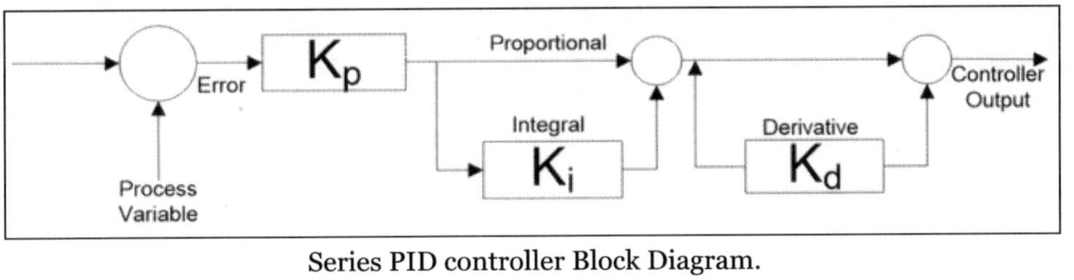

Series PID controller Block Diagram.

Applications of PID Controller

Temperature Control

Let us take an example of AC (air-conditioner) of any plant/process. Setpoint is temperature (20 °C) and current measured temperature by sensor is 28 °C. Our aim is to run AC at desired temperature (20 °C). Now, controller of AC, generate signal according to error (8 °C) and this signal is given to the AC. According to this signal, output of AC is changed and temperature decrease to 25 °C. further same process will repeat until temperature sensor measure desired temperature. When error is zero, controller will give stop command to AC and again temperature will increase up to certain value and again error will generate and same process repeated continuously.

Designing of MPPT (Maximum Power Point Tracking) Charge Controller for Solar PV

The I-V characteristic of a PV cell depends on temperature and irradiance level. So, operating voltage and current will change continuously with respect to change in atmospheric conditions. Therefore, it is very important to track maximum power point for an efficient PV system. To find MPPT, PID controller is used and for that current and voltage setpoint is given to the controller. If atmospheric conditions will change, this tracker keeps voltage and current constant.

Power Electronics Converter

PID controller is most useful in power electronics application like converters. If a converter is connected with system, according to change in load, output of converter must change. For example, an inverter is connected with load, if load increase more current will flow from inverter. So, voltage and current parameter is not fix, it will change according to requirement. In this condition, PID controller is used to generate PWM pulses for switching of IGBTs of inverter. According to change in load, feedback signal is given to controller and it will generate error. PWM pulses are generated according to error signal. So, in this condition we can get variable input and variable output with same inverter.

Supervisory Control and Data Acquisition

SCADA stands for supervisory control and data acquisition. It is a type of software application program for process control. SCADA is a central control system which consist of controllers network interfaces, input/output, communication equipments and software. SCADA systems are used to monitor and control the equipments in the industrial process which include manufacturing, production, development and fabrication. The infrastructural processes include gas and oil distribution, electrical power, water distribution. Public utilities include bus traffic system, airport. The SCADA system takes the reading of the meters and checks the status of sensors in regular interval so that it requires minimal interference of human.

A large number of processes occur in large industrial establishment. Every process you need to monitor is very complex because each machine gives different output. The

SCADA system used to gather the data from sensors and instruments located at remote area. The computer then processes this data and presents in a timely manner. The SCADA system gathers the information (like leak on a pipeline occurred) and transfer the information back to the system while giving the alerts that leakage has occurred and displays the information in a logical and organized fashion. The SCADA system used to run on DOS and UNIX operating systems.

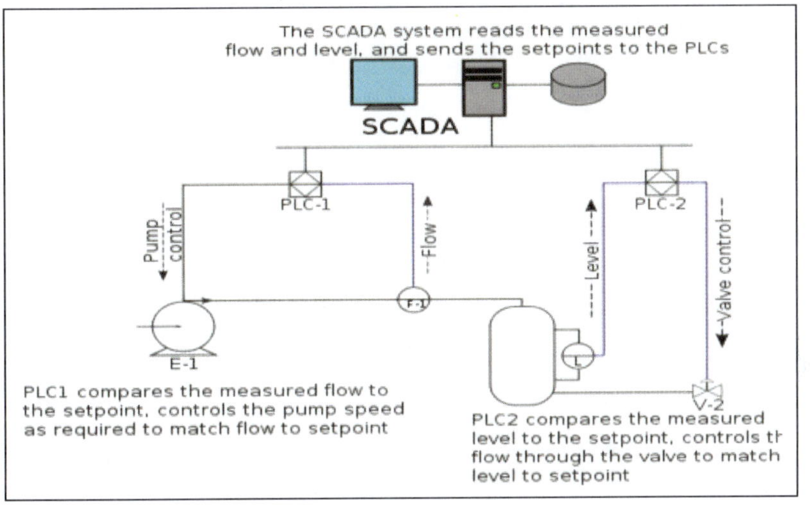

Architecture

Generally SCADA system is a centralized system which monitors and controls entire area. It is purely software package that is positioned on top of hardware. A supervisory system gathers data on the process and sends the commands control to the process. The SCADA is a remote terminal unit which is also known as RTU. Most control actions are automatically performed by RTUs or PLCs. The RTUs consist of programmable logic converter which can be set to specific requirement. For example, in the thermal power plant the water flow can be set to specific value or it can be changed according to the requirement.

The SCADA system allows operators to change the set point for the flow, and enable alarm conditions in case of loss of flow and high temperature and the condition is displayed and recorded. The SCADA system monitors the overall performance of the loop. The SCADA system is a centralized system to communicate with both wire and wireless technology to Clint devices. The SCADA system controls can run completely all kinds of industrial process. Example: If too much pressure in building up in a gas pipe line the SCADA system can automatically open a release valve.

Hardware Architecture

The generally SCADA system can be classified into two parts:

- Clint layer: The Clint layer which caters for the man machine interaction.

- Data server layer: The data server layer which handles most of the process data activities.

The SCADA station refers to the servers and it is composed of a single PC. The data servers communicate with devices in the field through process controllers like PLCs or RTUs. The PLCs are connected to the data servers either directly or via networks or buses. The SCADA system utilizes a WAN and LAN networks, the WAN and LAN consists of internet protocols used for communication between the master station and devices. The physical equipments like sensors connected to the PLCs or RTUs. The RTUs convert the sensor signals to digital data and sends digital data to master. According to the master feedback received by the RTU, it applies the electrical signal to relays. Most of the monitoring and control operations are performed by RTUs or PLCs as we can see in the figure.

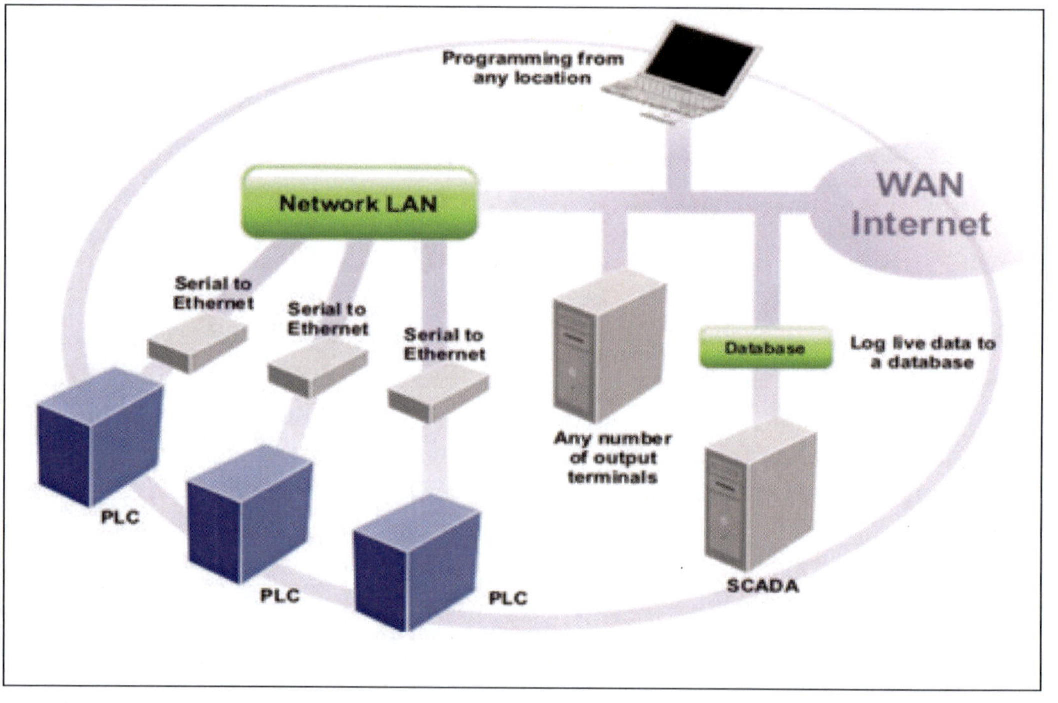

Software Architecture

Most of the servers are used for multitasking and real time database. The servers are responsible for data gathering and handling. The SCADA system consists of a software program to provide trending, diagnostic data, and manage information such as scheduled maintenance procedure, logistic information, detailed schematics for a particular sensor or machine and expert system troubleshooting guides. This means the operator can see a schematic representation of the plant being controlled. Example: Alarm checking, calculations, logging and archiving; polling controllers on a set of parameter, those are typically connected to the server.

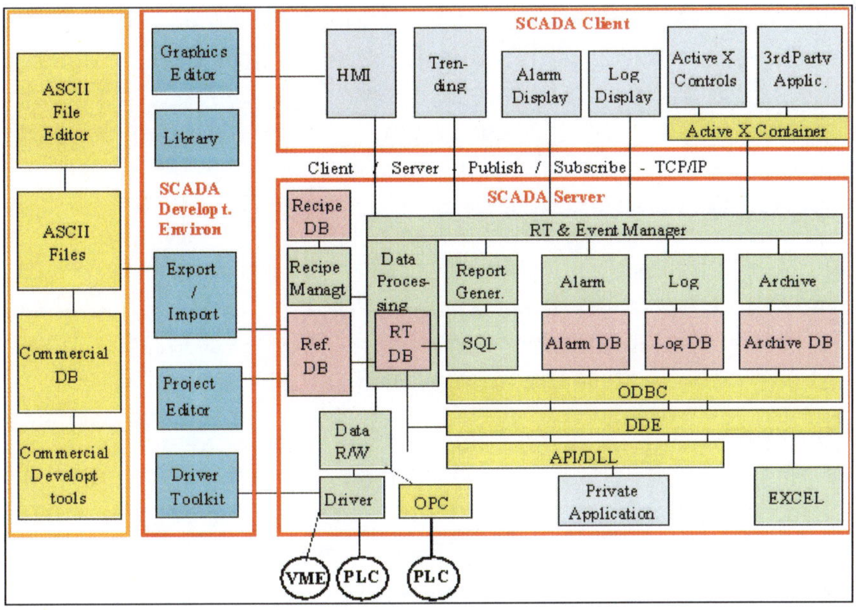

Typical generic software architecture of SCADA systems.

Working Procedure of SCADA System

The SCADA system performs the following functions:

- Data Acquisitions,
- Data Communication,
- Information/Data presentation,
- Monitoring/Control.

These functions are performed by sensors, RTUs, controller, communication network. The sensors are used to collect the important information and RTUs are used to send this information to controller and display the status of the system. According to the status of the system, the user can give command to other system components. This operation is done by the communication network.

Data Acquisitions

Real time system consists of thousands of components and sensors. It is very important to know the status of particular components and sensors. For example, some sensors measure the water flow from the reservoir to water tank and some sensors measure the value pressure as the water is release from the reservoir.

Data Communication

The SCADA system uses wired network to communicate between user and devices. The real time applications use lot of sensors and components which should be control remotely. The SCADA system uses internet communications. All information is transmitted through internet using specific protocols. Sensor and relays are not able to communicate with the network protocols so RTUs used to communicate sensors and network interface.

Information or Data Presentation

The normal circuit networks have some indicators which can be visible to control but in the real time SCADA system, there are thousands of sensors and alarm which are impossible to be handled simultaneously. The SCADA system uses human machine interface (HMI) to provide all of the information gathered from the various sensors.

The SCADA system uses human machine interface. The information is displayed and monitored to be processed by the human. HMI provides the access of multiple control units which can be PLCs and RTUs. The HMI provides the graphical presentation of the system. For example, it provides the graphical picture of the pump connected to the tank. The user can see the flow of the water and pressure of the water. The important part of the HMI is an alarm system which is activated according to the predefined values.

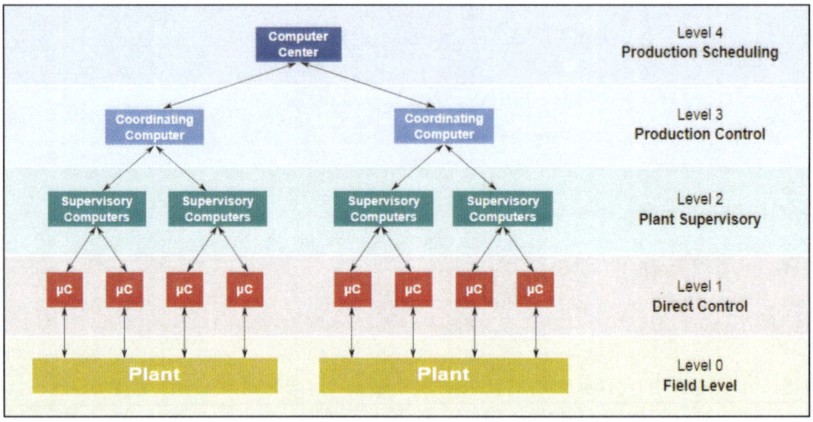

For example: The tank water level alarm is set 60% and 70% values. If the water level reaches above 60% the alarm gives normal warning and if the water levels reach above 70% the alarm gives critical warning.

Monitoring or Control

The SCADA system uses different switches to operate each device and displays the status at the control area. Any part of the process can be turned ON/OFF from the control station using these switches. SCADA system is implemented to work automatically without human intervention but at critical situations it is handled by man power.

SCDA for Remote Industrial Plant

In large industrial establishments many process occur simultaneously and each needs to be monitored, which is actually a complex task. The SCADA systems are used to monitor and control the equipments in the industrial processes which include water distribution, oil distribution and power distribution. The main aim of this project is to process the real time data and control the large scale remote industrial environment. In the real time scenario, a temperature logging system for a remote plant operation is taken.

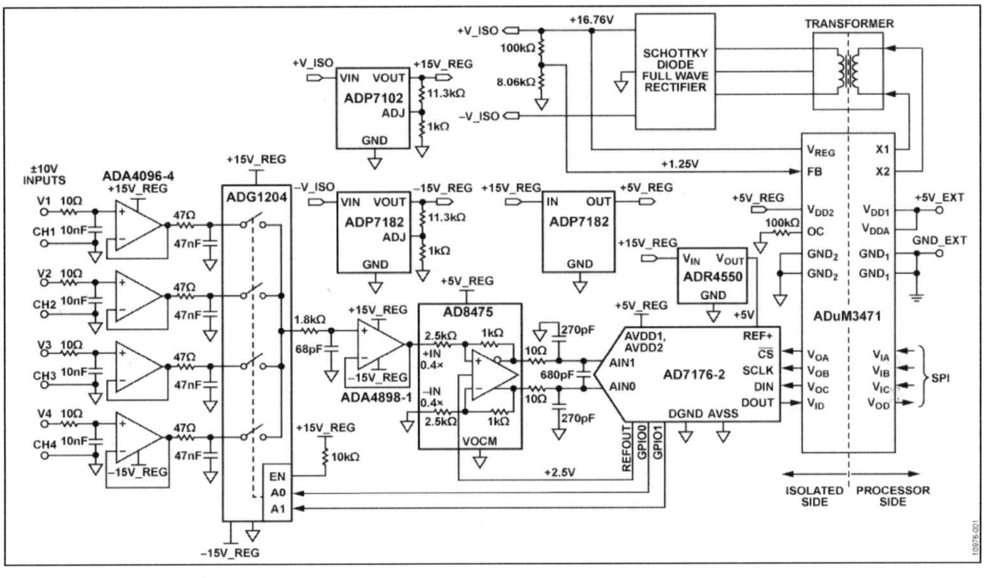

The temperature sensors are connected to the microcontroller, which is connected to the PC at the front end and software is loaded on the computer. The data is collected from the temperature sensors. The temperature sensors continuously send the signal to the microcontroller which accordingly displays these values on its front panel. One can set the parameters like low limit and high limit on the computer screen. When the temperature of a sensor goes above set point the microcontroller send a command to the corresponding relay. The heaters connected through relay contacts are turned OFF and ON.

For example SCADA for Remote Industrial Planet:

This is a temperature logging System. Here 8 temperature sensors in multiplexing mode are connected to the microcontroller through ADC 0808. Then the values of all the sensors are sent serially by microcontroller through Max 32 to the com port of the PC. Software "DAQ System" loaded on the PC takes these values and shows them on its front panel, and also logs them to the data base "daq.mdb". One can set by interactive way some parameters like set point, low limit, and high limit on the computer screen. When temperature of some sensor increases beyond set point, the microcontroller sends commands to relay driver IC. The heaters connected through relay contacts are (specific for that sensor) turned OFF (or ON in opposite case). High limit and low limits are for alarm. When temperature goes above high limit or below low limit the alarm will be turned on.

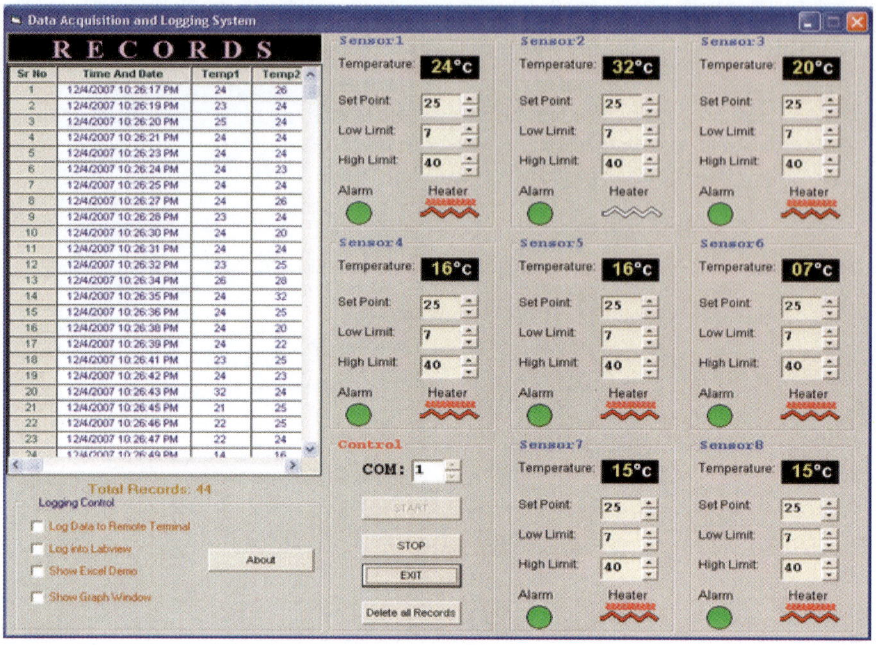

Applications

- Power generation, transmission and distribution.
- Water distribution and reservoir system.
- Public buildings like electrical heating and cooling system.
- Generators and turbines.
- Traffic light control system.

Advantages

- The SCADA system provides onboard mechanical and graphical information.
- The SCADA system is easily expandable. We can add set of control units and sensors according to the requirement.
- The SCADA system ability to operate critical situations.

SCADA Security Issues

Security of SCADA-based systems is being questioned, as they are potential targets to cyberterrorism/cyberwarfare attacks.

There is an erroneous belief that SCADA networks are safe enough because they are secured physically. It is also wrongly believed that SCADA networks are safe enough because they are disconnected from the Internet.

SCADA systems also are used for monitoring and controlling physical processes, like distribution of water, traffic lights, electricity transmissions, gas transportation and oil pipelines and other systems used in the modern society. Security is extremely important because destruction of the systems would have very bad consequences.

There are two major threats. The first one is unauthorized access to software, be it human access or intentionally induced changes, virus infections or other problems that can affect the control host machine. The second threat is related to the packet access to network segments that host SCADA devices. In numerous cases, there remains less or no security on actual packet control protocol; therefore, any person sending packets to SCADA device is in position to control it. Often, SCADA users infer that VPN is sufficient protection, and remain oblivious to the fact that physical access to network switches and jacks related to SCADA provides the capacity to bypass the security on control software and control SCADA networks.

SCADA vendors are addressing these risks by developing specialized industrial VPN and firewall solutions for SCADA networks that are based on TCP/IP. Also, white-listing solutions have been implemented due to their ability to prevent unauthorized application changes.

Programmable Logic Controller

Programmable Logic Controller (PLC) focuses on automating different electro-mechanical processes taking place in various industries. With a well-programmed microprocessor, It has specially designed and implemented controllers which are immune to extremely high and low temperatures, humidity, dust, etc.

Just like any usual program, the coding is programmed on a computer. From here it is transferred to the PLC using a cable. After downloading the programs, these are loaded and stored in the PLC. In order to make sure the programs do not vanish at the time of power cut or switch off, the PLC uses non-volatile memory storage.

It is programmed using a programming language called Ladder Logic. It is already known that PLC is an industrial computer. In order to program a PLC, a programming language had to be designed in such a way, that the electricians and electrical engineer in an industry could easily understand and feel comfortable enough to code in it.

A Programmable Logic Controller consists of various input and output terminals. Using the input terminals, PLC interprets the logical states from switches and sensors. There are two logical states also known as high(or 1) and low(or 0). The output from the output terminals is used as a signal for various devices, providing them on/off controls.

The electrical engineers understand the ladder logic diagrams and so it was the inspiration behind the programming language used to program its.

The most commons areas of use of PLCs is in a washing machine, elevators and also traffic signals.

Programmable Logic Controller, more popularly known as PLC is a digital computer without a mouse or a keyboard or a monitor.

Programmable Logic Controller is a small computer which automatically controls various processes and components in an industrial system. For the sake of comparison, let's take a process that is manual as an example thinks of PLC as a brain controlling a finger on a trigger of an applicator. The brain sends a signal to the finger directing it to pull the finger when something is dispensed and when the brain wants to stop it sends another signal. Like most of the brains, PLC can multitask and it's lightning fast Just like that, it can control much input and output signal like direct a paint line to change colors so all the widgets being painted go from one color to other.

PLC Architecture

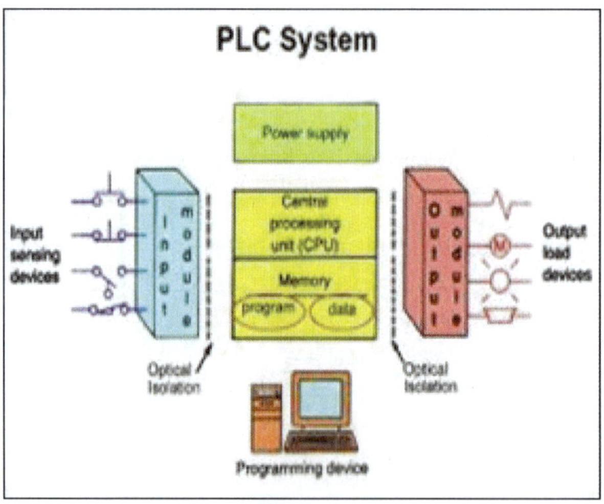

PLC Internal Architecture.

A basic PLC system consists of the following:

- Input/Output Section: The input section or input module consists of devices like sensors, switches and many other real world input sources. The input from the sources is connected to the PLC through the input connector rails. The output section or output module can be a motor or a solenoid or a lamp or a heater, whose functioning is controlled by varying the input signals.
- CPU or Central Processing Unit: It is the brain of the PLC. It can be a hexagonal or an octal microprocessor. It carries out all the processing related to the input signals in order to control the output signals based on the control program.

- Programming Device: It is the platform where the program or the control logic is written. It can be a handheld device or a laptop or a computer itself.
- Power Supply: It generally works on a power supply of about 24 V, used to power input and output devices.
- Memory: The memory is divided into two parts- The data memory and the program memory. The program information or the control logic is stored in the user memory or the program memory from where the CPU fetches the program instructions. Thc input and output signals and the timer and counter signals are stored in the input and output external image memory respectively.

Working of a PLC

- The input sources convert the real time analog electric signals to suitable digital electric signals and these signals are applied to the PLC through the connector rails.
- These input signals are stored in the PLC external image memory in locations known as bits. This is done by the CPU.
- The control logic or the program instructions are written onto the programming device through symbols or through mnemonics and stored in the user memory.
- The CPU fetches these instructions from the user memory and executes the input signals by manipulating, computing, processing them to control the output devices.
- The execution results are then stored in the external image memory which controls the output drives.
- The CPU also keeps a check on the output signals and keeps updating the contents of the input image memory according to the changes in the output memory.
- The CPU also performs internal programming functioning like setting and resetting of the timer, checking the user memory.

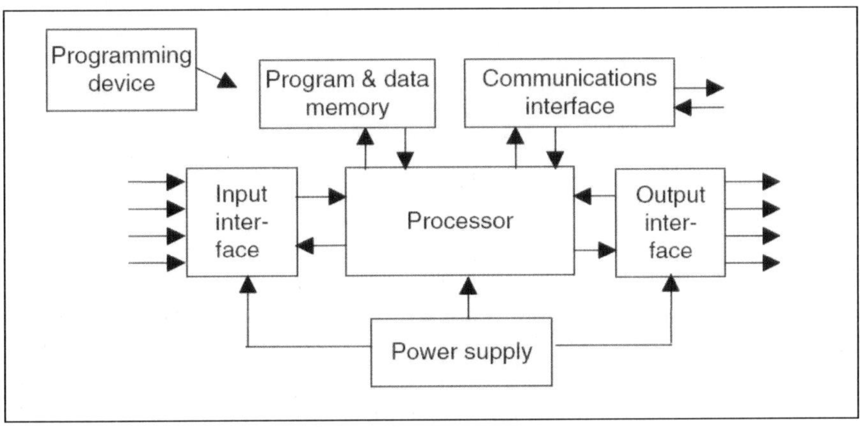

PLC Working Schematic.

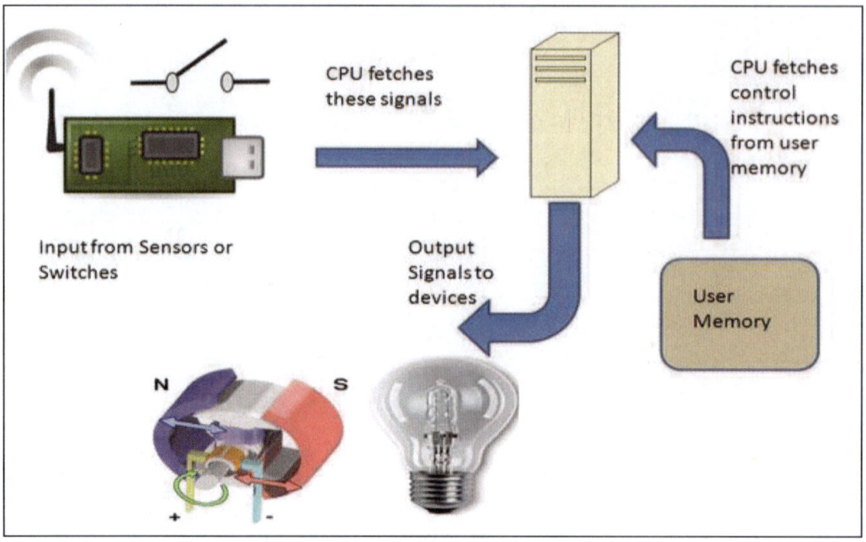

Working of PLC.

Programming in PLC

The basic functioning of the PLC relies on the control logic or the programming technique used. Programming can be done using flowcharts or using ladder logic or using statement logics or mnemonics.

Interlinking all these, let us see how we can actually write a program in PLC.

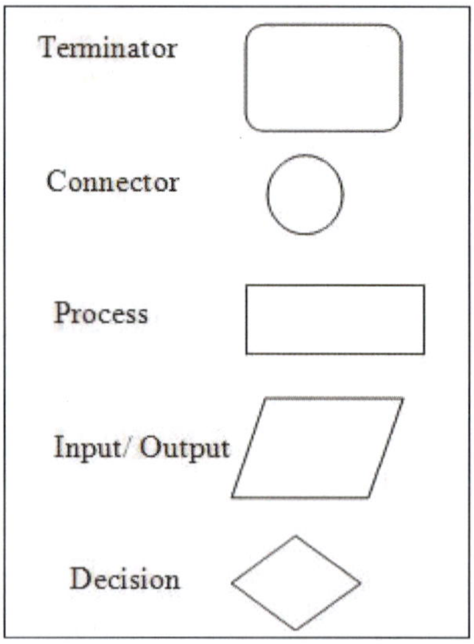

- Write the Boolean expression for the different logic. Boolean algebra usually involves logic operations like AND, OR, NOT, NAND and NOR. The different symbols are:

  + OR operator,

. AND operator,

! NOT operator.

- Write the instructions in simple statement forms like: IF Input1 AND Input2 Then SET Output1 ELSE SET Output.
- Write the ladder logic program. It is the most important part of PLC programming. Before explaining about ladder logic programming, let us know about few symbols and terminologies.
- Rung: One step in the ladder is called a rung. In simpler words, the basic statement or one control logic is called a Rung.

Y: Normal Output signals,

M: Motor symbol,

T: Timer,

C: Counter.

Symbols:

<table>
<tr><td>| |</td><td>Load symbol or switch in ON condition.</td></tr>
<tr><td>|/|</td><td>Load Inverse or switch in OFF condition</td></tr>
<tr><td>( )</td><td>Output Symbol</td></tr>
<tr><td>[CJP]</td><td>Conditional jump</td></tr>
<tr><td>[EJB]</td><td>End jump</td></tr>
<tr><td>[SFT]</td><td>Shift</td></tr>
</table>

Basic Logic Functions using Ladder Logic:

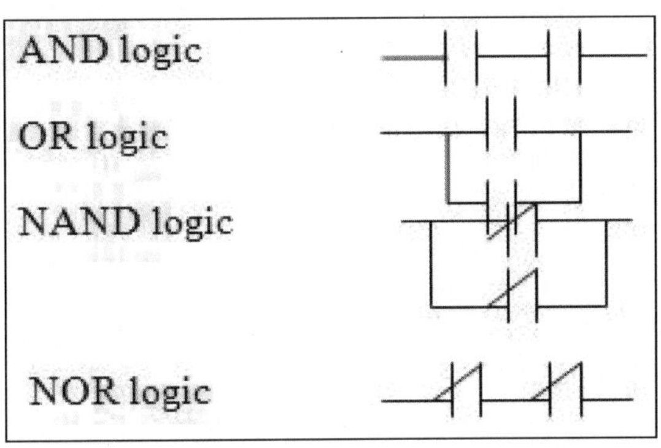

- Writing Mnemonics: Mnemonics are instructions written in symbolic form. They are also known as Opcode and are used in handheld programming devices. Different Symbols are as given below:

  Ldi: Load Inverse,

  Ld: Load,

  AND: And logic,

  OR: Or logic,

  ANI: NAND logic,

  ORI: NOR logic,

  Out: Output.

A Simple PLC Application

So, now that we have had a brief idea about programming in PLC, let's get into developing one simple application.

Problem: Design a simple line follower robotic system to start a motor when a switch is on and simultaneously switch on the LED. The sensor on the motor detects any obstacle and another switch is on to indicate the presence of the obstacle and the motor is simultaneously switched off and the buzzer is switched on and LED is off.

Solution:

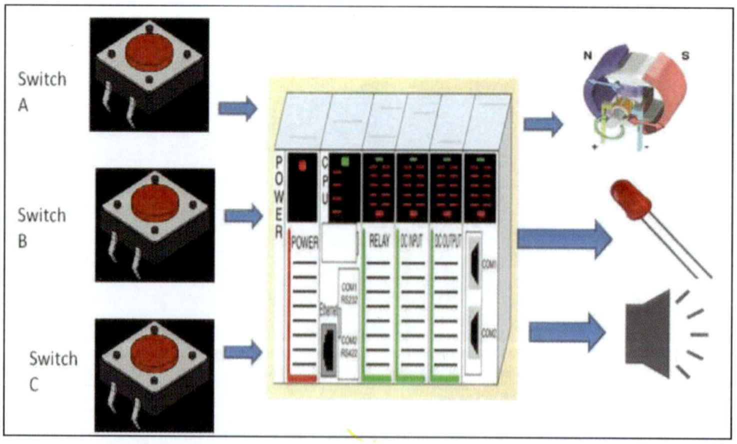

Let us first assign our symbols or tags to the inputs and outputs:

M – Motor,

A – Input Switch 1,

B- Input Switch 2,

L – LED,

Bu –Buzzer.

Now let us design the Flow Chart:

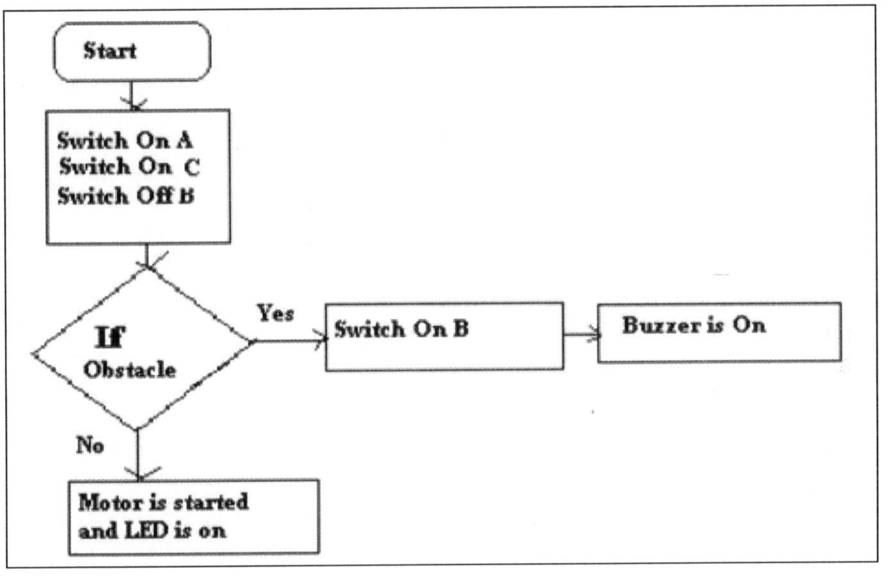

Next Step is writing the Boolean expressions:

M = A. (! B),

L = C. (! B),

Bu = B. (! A.! C).

The next step involves drawing the ladder logic program:

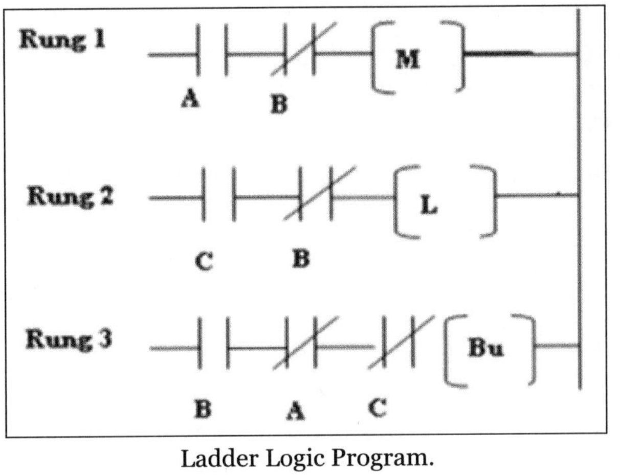

Ladder Logic Program.

The final step involves writing the mnemonics to be fed to the handheld device:

Ld A ANI Ldi B,

Ld C ANI Ldi B,

Ld B ANI Ldi A AND Ldi C.

Human Machine Interface

HMI is the acronym for Human Machine Interface, and can be designed as just that; an interface between the user and the machine. An HMI is considered an interface; a very broad term that can include MP3 players, industrial computers, household appliances, and office equipment. However, an HMI is much more specific to manufacturing and process control systems. An HMI provides a visual representation of a control system and provides real time data acquisition. An HMI can increase productivity by having a centralized control center that is extremely user-friendly.

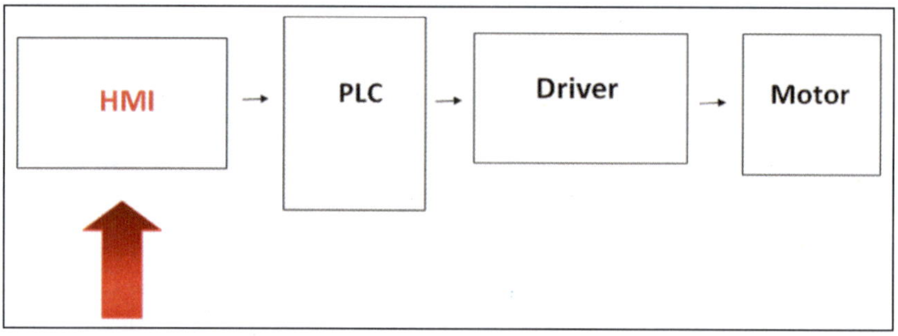

HMI Block Diagram.

A Human Machine Interface (HMI) is exactly what the name implies; a graphical interface that allows humans and machines to interact. Human machine interfaces vary widely, from control panels for nuclear power plants, to the screen on an iPhone. However, for this discussion we are referring to an HMI control panel for manufacturing-type processes. An HMI is the centralized control unit for manufacturing lines, equipped with Data Recipes, event logging, video feed, and event triggering, so that one may access the system at any moment for any purpose. For a manufacturing line to be integrated with an HMI, it must first be working with a Programmable Logic Controller (PLC). It is the PLC that takes the information from the sensors, and transforms it to Boolean algebra, so the HMI can decipher and make decisions.

Basic Types of HMIs

There are three basic types of HMIs: the pushbutton replacer, the data handler, and the overseer. Before the HMI came into existence, a control might consist of hundreds of pushbuttons and LEDs performing different operations. The pushbutton replacer HMI has streamlined manufacturing processes, centralizing all the functions of each button into one location. The data handler is perfect for applications requiring constant feedback from the system, or printouts of the production reports. With the data handler, you must ensure the HMI screen is big enough for such things as graphs, visual representations and production summaries. The data handler includes such functions as recipes, data trending, data logging and alarm handling/logging. Finally, anytime an application involves SCADA or MES, an overseer HMI is extremely beneficial. The overseer HMI will most likely need to run Windows, and have several Ethernet ports.

An HMI is a substantial purchase, so it is important to know exactly what is required of it. An HMI is used for three primary roles: a pushbutton replacer, data handler, and overseer. The pushbutton replacer takes the place of LEDs, On/Off buttons, switches or any mechanical device that performs a control function. The elimination of these mechanical devices is possible because the HMI can provide a visual representation of all these devices on its LCD screen, while performing all the same functions. The Data Handler is used for applications that require constant feedback and monitoring. Often these Data Handlers come equipped with large capacity memories. The last of the HMI three types is referred to as the overseer, because it works with SCADA and MES. These are centralized systems that monitor and control entire sites or complexes of large systems spread out over large areas. An HMI is usually linked to the SCADA system's databases and software programs, to provide trending, diagnostic data, and management information.

Physical Properties of a HMI

The actual physical properties of an HMI vary from model to model and among manufacturers. It is important that one makes the appropriate selection. An HMI that is located in a water plant might have various water seals around its perimeter, as opposed to an HMI that is located in a pharmaceutical warehouse. The actual size of an HMI is also a key physical property that will vary, because not all applications require a large, high-resolution monitor. Some applications may only require a small, black and white touch screen monitor. When it comes to selecting an HMI, the physical properties are extremely important because one must take into consideration the operating environment, and what safety measures the HMI needs to protect itself. Also, a specific size may be needed due to space limitations. Lastly, physical properties include the processor and memory of the HMI. It is important to make sure that the processor and memory capabilities are sufficient enough to control a system.

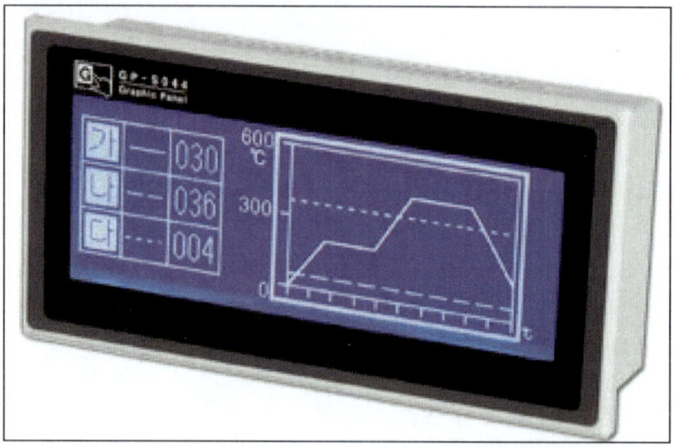

Working of HMI

First consider the other components that are necessary to make a manufacturing control system operate. The production line consists of all the machinery that performs the work required in the production of the product. Next, consider the various input/output sensors that monitor temperature, speed, pressure, weight and feed rate. Third, decide on the programmable logic controller (PLC) that will receive the data from the input/output sensors, and converts the data into logical combinations.

Environmental Aspects for an HMI

The operating environment should always be considered when selecting an HMI. For example: for use in a warehouse that has excessive noise or vibration, one might decide on a heavy-duty HMI. If used in the food processing industry or somewhere that might need to be washed down, select a water-protected HMI. One must also consider temperature as a factor in the selection process. If used in a steel plant, next to a furnace, one would want something that can withstand extreme temperatures.

Choosing a Programming Software

When considering which programming software to use, there are three main categories to choose from: proprietary, hardware independent and open software. Proprietary software is the software that the manufacturer provides, which is typically easy to use and allows for quick development. The drawback is that proprietary software will only run on that specific hardware platform. Hardware-independent software is third party software developed to program on several different types of HMIs. This type of software gives the developer much more freedom for the HMI selection. The downside to hardware independent software is that it is not as user-friendly as the proprietary. Open Software should only be selected by the advanced programmer. It allows the developer to have complete openness in the design process.

HMI Applications

The HMI is used throughout various industries including manufacturing plants, vending machines, food and beverage, pharmaceuticals, and utilities, just to name a few. HMIs along with PLCs are typically the backbone of the production line in these industries. The integration of the HMI into manufacturing has vastly improved operations. The HMI allows for supervisory control and data acquisition in the entire system, so parameter changes are feasible as the operator's choosing. For example, in metals manufacturing, an HMI might control how metal is cut and folded, and how fast to do so. An HMI offers improved stock control and replenishment, so the fewer journeys are required out to the vendors. HMIs are used in bottling processes to control all aspects of the manufacturing line, such as speed, efficiency, error detection and error correction. Utility companies may use HMIs to monitor water distribution and waste water treatment.

Advantages of an HMI

The greatest advantage of an HMI is the user-friendliness of the graphical interface. The graphical interface contains color coding that allows for easy identification (for example: red for trouble). Pictures and icons allow for fast recognition, easing the problems of illiteracy. HMI can reduce the cost of product manufacturing, and potentially increase profit margins and lower production costs. HMI devices are now extremely innovative and capable of higher capacity and more interactive, elaborate functions than ever before. Some technological advantages the HMI offers are: converting hardware to software, eliminating the need for mouse and keyboard, and allowing kinesthetic computer/human interaction.

HMIs communicate with Programmable Logic Controllers (PLCs) and input/output sensors to get and display information for users to view. HMI screens can be used for a single function, like monitoring and tracking, or for performing more sophisticated operations, like switching machines off or increasing production speed, depending on how they are implemented.

HMIs are used to optimize an industrial process by digitizing and centralizing data for a viewer. By leveraging HMI, operators can see important information displayed in graphs, charts, or digital dashboards, view and manage alarms, and connect with SCADA and MES systems, all through one console.

Previously, operators would need to walk the floor constantly to review mechanical progress and record it on a piece of paper or a whiteboard. By allowing PLCs to communicate real-time information straight to an HMI display, HMI technology eliminates the need for this outdated practice and thereby reduces many costly problems caused by lack of information or human error.

Difference between HMI and SCADA

Supervisory Control and Data Acquisition (SCADA) and HMI are closely related, and often referred to in the same context since they are both part of a larger industrial control system, but they each offer different functionality and opportunities. While HMIs are focused on visually conveying information to help the user supervise an industrial process, SCADA systems have a greater capacity for data collection and control-system operation. Unlike SCADA systems, HMIs do not collect and record information or connect to databases. Rather, HMI provides an effective communication tool that functions as part of, or alongside, a SCADA system.

Developing Trends in HMI Technology

In the past decade, changing operational and business needs have instigated interesting developments in HMI technology. Now, it's becoming more common to see evolved forms of HMI such as high-performance HMIs, touch screens, and mobile devices, along with more traditional models. These modernized interfaces are creating more opportunities for equipment interaction and analysis.

High-performance HMIs

Operators and users are increasingly moving toward high-performance HMI, a method of HMI design that helps ensure fast, effective interaction. By only drawing attention to the most necessary or critical indicators on the interface, this design technique helps the viewer to see and respond to problems more efficiently, as well as make better-informed decisions. Indicators on high-performance HMI are simple, clean, and purposely cleared of any extraneous graphics or controls. Other design elements, like color, size, and placement, are used with discretion to optimize the user experience.

Touch Screens and Mobile Devices

Touch screens and mobile HMI are two examples of technological advances that have emerged with the advent of smartphones. Instead of buttons and switches, modernized HMIs allow operators to tap or touch the physical screen to access controls. Touch screens are especially important when used with mobile HMI, which is either deployed through web-based HMI/SCADA or via an application. Mobile HMI offers a variety of advantages to operators, including instant access to HMI information and remote monitoring.

Remote Monitoring

Mobile-friendly remote monitoring allows greater flexibility and accessibility for operators and managers alike. With this feature, an offsite control system engineer can, for example, confirm the temperature of a warehouse on a portable device, eliminating the need for onsite supervision after working hours. Soon, checking in on a process on your factory floor while being miles away from the facility won't seem like anything out of the ordinary.

Edge-of-network and Cloud HMIs

Edge-of-network HMIs are also in high demand because they allow operators to access data and visualization from field devices. Additionally, it is becoming more common to send data from local HMIs to the cloud, where it can be accessed and analyzed remotely, while keeping control capabilities local.

Artificial Neural Network

Artificial Neural Networks are software implementations of the neural structures of human brain. ANN is a computational system influenced from the structure, processing capability and learning ability of a human brain.

Not diving deep into the complex biology of it, let us take a look at the structure of our brain. Human brain contains billions of neurons that act as organic switches. All these neurons are interconnected to form a huge and complex structure called Neural

Network. The output of a single neuron is dependent on inputs from thousands of interconnected neurons.

The "Learning" of a human brain is simply repeated activation of certain neural connections and this repetition strengthens the connection. So, for a specified input, the neural connections make sure that output is always a desired one. A simple feedback from the outcome helps the learning process as it strengthens the neural connections.

This behavior of the brain is key to Artificial Neural Networks as they simply try to replicate this action of the brain. This can be achieved in two ways:

- Supervised ANN,
- Unsupervised ANN.

Supervised Artificial Neural Networks

In Supervised Artificial Neural Networks, a matched input and output samples of data is provided to the network for training. The aim of this approach is to get a desired output for a specified input.

One of the best examples of a Supervised ANN is the spam filters of our e-mails. At training level, the input to the Artificial Neural Network engine would be a set of words in the body of the e-mail. The output is flagging the e-mail as either spam or not spam.

Unsupervised Artificial Neural Networks

The Unsupervised Artificial Neural Network is more complex than the supervised counterpart as it attempts to make the ANN understand the data structure provided as input on its own.

Characteristics of Artificial Neural Networks

Any Artificial Neural Network, irrespective of the style and logic of implementation, has a few basic characteristics:

- An Artificial Neural Network consists of large number of "neuron" like processing elements.
- All these processing elements have a large number of weighted connections between them.
- The connections between the elements provide a distributed representation of data.
- A Learning Process is implemented to acquire knowledge.

Elements of Artificial Neural Networks

Keeping the above characteristics in mind, we can derive the basic elements of any Artificial Neural Network as follows:

- Processing Elements,
- Topology,
- Learning Algorithm.

Processing Elements

As the ANN is a simplified computational model of a biological neural network, an ANN consists of basic processing units or elements similar to that of neurons of a brain.

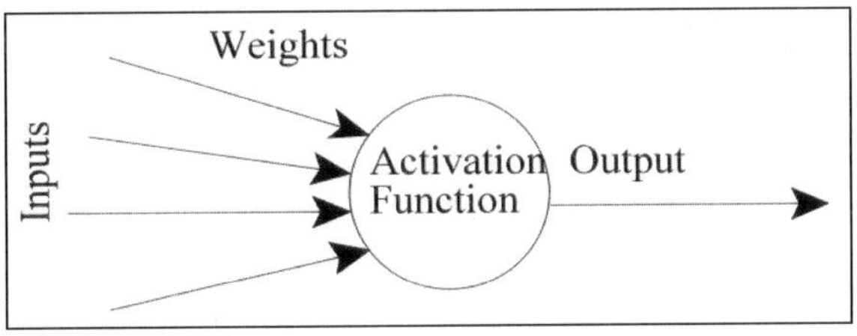

In general, a processing unit is made up of summing unit followed by an output unit. The function of a summing unit is to take n input values, weight each input value and calculate the weighted sum of those values.

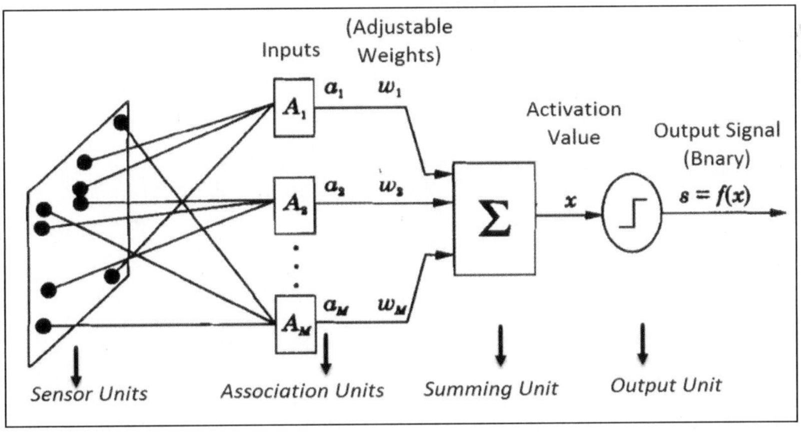

Based on the sign of the weight of each input, it is determined whether the input has a positive weight or a negative weight. The weighted sum of the summing unit is known as Activation Value and based on the signal from this activation value, the output is produced.

Both the input and output can be either continuous or discrete as well as they can be either deterministic or fuzzy.

Topology

Any Artificial Neural Network will become useful only when all the processing elements are organized in an appropriate manner so that they can accomplish the task of pattern recognition.

This organization or arrangement of the processing elements, their interconnections, inputs and outputs is simply known as Topology.

Generally, in an ANN, the processing units are arranged into layers and all the units in a particular layer have the same activation values and output values. Connection can be made between layers in multiple ways like processing unit of one layer connected to a unit of another layer, processing unit of a layer connected to a unit of same layer, etc.

Some of the commonly used Topologies in Artificial Neural Networks are:

- Instar,
- Outstar,
- Group of Instars,
- Group of Outstars,
- Bidirectional Associative Memory,
- Autoassociative Memory.

The following image shows an arrangement of two layers F1 and F2 with M and N number of processing units each in both Instar and Outstar Topologies.

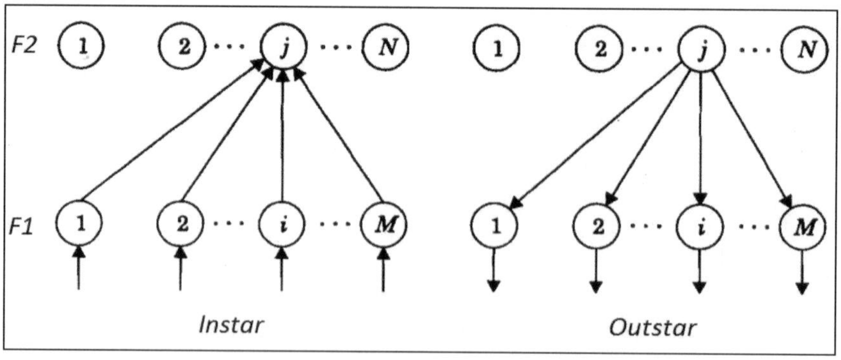

Learning Algorithm

The final and important elements of any ANN are Learning Algorithms or Laws. The operation any neural network is governed by Neural Dynamics consisting of both activation state dynamics and synaptic weight dynamics.

Learning Algorithms or Laws are implementations of synaptic dynamics and are described in terms of first derivative of the weights. These leaning laws can be supervised, unsupervised or a hybrid of both.

Some of the commonly known Learning Algorithms are:

- Hebb's Law,
- Perception Learning Law,
- Delta Learning Law,
- Wildrow & Hoff LMS Learning Law,
- Correlation Learning Law,
- Instar Learning Law,
- Outstar Learning Law.

Different Types of Artificial Neural Networks

There are many different flavors of Artificial Neural Networks. Some of them are listed here:

- Feedforward Neural Networks,
- Feedback Neural Networks,
- Competitive Learning Neural Networks.

Feedforward Neural Networks

A Feedforward Artificial Neural Network, as the name suggests, consists of several layers of processing units where each layer is feeding input to the next layer, in a feed-through manner. A simple two-layer network is an example of feedforward ANN.

The following is a simple structure of a three-layered feedforward ANN:

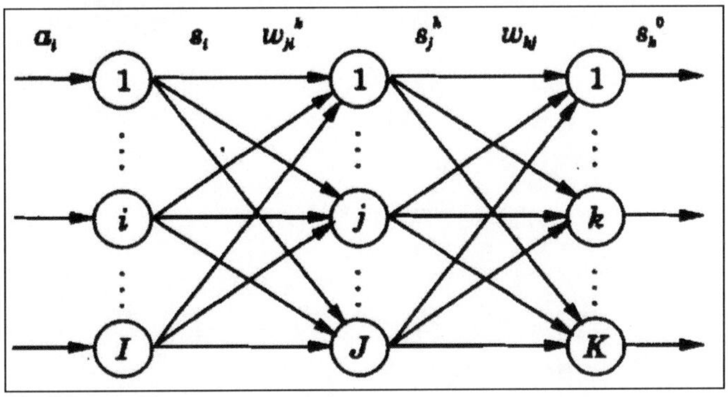

Feedback Neural Networks

A feedback ANN consists of the feedback element. In a basic structure, a feedback ANN consists of a set of processing units, whose outputs are fed back as inputs to all other units of the same layer including the same unit.

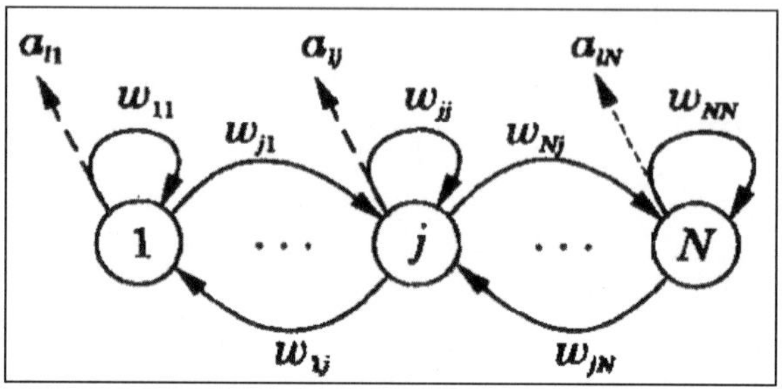

In its basic form, a feedback ANN doesn't have any structure and hence it is not an useful at pattern recognition.

Competitive Learning Neural Networks

It is a combination of both feedback and feedforward ANNs. The input layer is linear and its outputs are given to all the units in the next layer.

The outputs of the second layer can be either linear or non-linear (depending on the application).

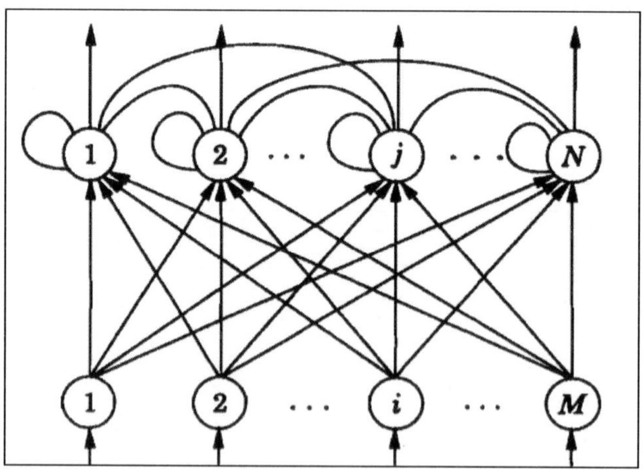

Advantages

- The main advantage of ANN is parallel processing. This makes it more useful that linear program.
- Due to their parallel processing structure, any failure in one neural element will not affect the rest of the process.

- Neural networks can be applied to any application and they can solve any complex problem.
- By implementing appropriate learning algorithms, an ANN can be made to learn without reprogramming.

Disadvantages

- All the parallel processing requires a huge amount of processing power and time.
- There is a requirement for a "training" period before real-world implementation.

Applications of ANN

The two important areas where ANNs have a huge potential of applications are Speech and Image Processing:

- Applications in Speech:
  - Vowel Classification.
  - Recognition of vowel-consonant segments.
  - Recognition of stop consonant-vowel utterances in Indian languages.
  - Nettalk.
  - Phonetic Typewriter.
- Applications in Image Processing:
  - Recognition of Symbols (used in Olympics).
  - Recognition of handwriting.
  - Segmentation of image.
  - Classification and segmentation of texture.

ANN Modelling to Optimize Manufacturing Process

The use of artificial intelligence and specifically artificial neural networks (ANNs) has allowed yielding revolutionary advances in manufacturing. However, most of the applications of artificial intelligence in the production field concerned expert systems and fewer attentions were paid to neural networks (NNs). Most important characteristics of the ANNs are:

- The self-adaptive behavior that allows to adapt the forecast to changing of the environment, in this way improve the networks' ability to learn and to predict.

- The parallel computing architecture, that has a great impact in multiple disciplines and applications, from speech and natural language processing, to image processing or problems in bioinformatics and biomedical engineering.

Therefore, they could be of great help for today's computer integrated manufacturing and in smart factories, according to Industry 4.0 paradigm. Currently, the nature of the manufacturing process is changing with great speed, becoming more sophisticated and continuous variations are occurring due to changes in customer demand and reduced product life cycle. This requires manufacturing technologies that can easily adapt to such changes. In this context, artificial neural networks are a powerful technology to solve this problem. The use of ANNs is also widely used for process monitoring and control applications. The quality of a process can only be provided by in process monitoring through proper measurements. To ensure a high quality of a process, you must follow the following technological steps:

- Identify the characteristic changes of a process.
- To estimate the changes of the product quality.
- Correct any process operations as a result of any anomalies detected from the comparison between the obtained and the desired quality.

These steps should be followed with minimal supervision and assistance from operators, if possible in unmanned manner. In addition, all processes should be implemented with special features such as: storing information, decision making, learning and integration. It should be noted that most manufacturing processes are regulated by many variable parameters and for this reason; such systems have a random, complex and uncertain nature.

This may be attributable to the fact that they are exposed to external disturbance and noise and often subjected to parameter variations. Furthermore, there is often a great interaction between variables and therefore it is not possible to properly define the final quality of the product and the variables that influence it. Due to these characteristics the quality often varies from product to product, impairing its uniformity and decreasing the yield of the product. So, all changes that may occur in manufacturing environments cannot be easily observed by an operator, so in recent years the use of neural networks applied to process monitoring and control has been of great interest. Indeed, it has been shown that the use of artificial intelligence can overcome the above-mentioned problems. The research efforts in this direction will be accelerated with greater interest in the future and will lead to the development of truly intelligent manufacturing systems that are capable of producing products without the supervision or assistance of human operators. Figure classifies the functionalities needed to imbed the artificial neural networks on manufacturing processes and summarizes the current developments in manufacturing application areas.

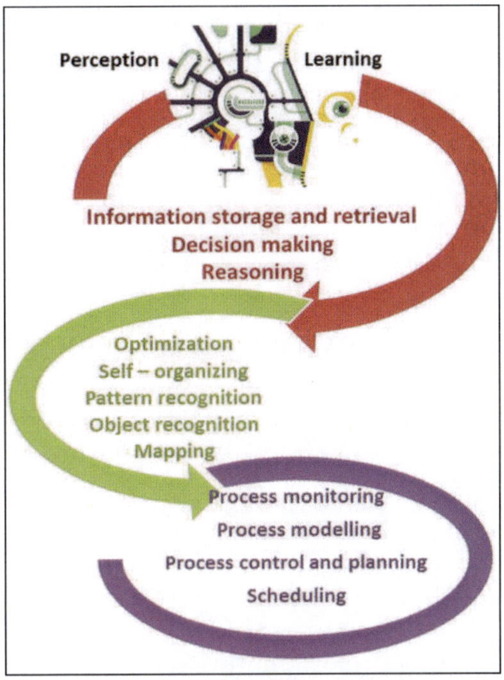

Functionality of the artificial neural networks and their manufacturing applications.

The real world applications here in manufacturing include the modelling, monitoring and control, identification, planning and scheduling associated with the processes.

Manufacturing Applications

Due to many external disturbances and many variations in process parameters, many production processes are complex and time-consuming. For these reasons, it is not always possible to identify the relationship between the product quality and the input variables of the process. Thus, there is interest to integrate the artificial intelligence into the production processes for storing, learning, reasoning and decision making. Such systems are able to adapt to changes in its environment and can truly realize unmanned operations of processes. The adoption of the neural network can be devoted to monitoring and to prediction of different parameters in many industrial areas, in order to solve issue relating to the manufacturing system design, process planning, as well as operational decision making.

Table: Summary of neural network manufacturing applications.

Category	Manufacturing
Application topic	• Manufacturing system design
	• Manufacturing process control
	• Robot scheduling
	• Manufacturing operational decision

Manufacturing information, such as the sequence of operations, lot size; multiple process plans were given special consideration in their approach to solve the generalized part family formation problem.

Injection Moulding Processes

Injection moulding processes are characterized by dynamic characteristics since process input variables are the melting temperatures, the velocity of the cylinder, the holding, the pressure that produces the polymer flow into the model cavity and they vary in a complex manner. The phenomena occurring in the process are very complex, time-varying, nonlinear and uncertain. This complexity makes it difficult to relate the input operating variables to the product quality such as geometry accuracy and geometry surface smoothness. These processes have been implemented and optimized with the use of artificial intelligence with the use of multilayer perceptron which is found to be the most popular network and tries to model the process dynamics and based on this to predict the part quality.

Gas Metal Arc Welding Processes

In gas metal arc (GMA) welding processes, the flow of an electric current is generated by an electric arc that is maintained between the consumable wire electrode and the welding metal as shown in figure.

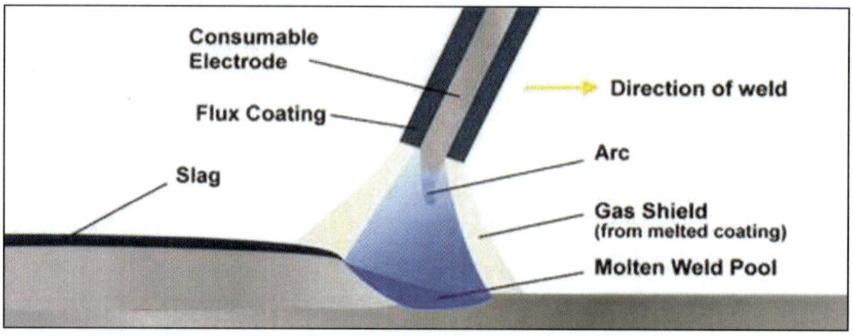

Characteristics of the GMA welding process.

Both the filling metal and the consumable electrode are automatically fed by a wire feeding device. A good quality of the welds is determined by the relatively high depth to width ratio of the molten welding pool. So, the monitoring and control of weld geometry and the surface temperatures that are strongly related with the formation of the weld pool, are very important for the penetration depth or the back bead width. For this purpose, the temperatures by noncontact were measured. Infrared temperature sensing system and recent studies conducted the ANN multilayer perceptron to detect and control with great success all the surface temperature information.

Arc Welding Processes

The complexity of the relationship between the process variable and weld quality is the common factor in all manufacturing processes and in particular arc welding processes. For these reasons, the use of ANNs for quality monitoring and control of the process. In this type of welding process, the ANN input data are generally the surface temperature, the welding voltage and the current and torch speed. The majority type of ANN employed for this case was, again, the multilayer perceptron. Their use was found to be very satisfactory to predict the weld defects, the geometry such as bead width, head height and penetration.

Machining Processes

To perform a correct quality control in all machining systems, it is very important to check particular parameters such as the cutting tool state, the vibrations, the forces and the temperature obtained during real-time machining operation.

Semiconductor Manufacturing Processes

The complexity of plasma etching processes in integrated circuits fabrication promoted the use of ANNs for monitoring and control. In this field, the use of artificial intelligence brings advantages that could not be achieved with traditional open loop controls. Where used the multilayer perceptron networks that are the most popular for this process. The ANNs use the fundamental parameters that affecting process dynamics, such as power, gas flow rate, dc bias voltage and throttle positioning. Proper use of networks in this field allows real-time monitoring and estimation of quality variables such as the etching thickness and the etching time.

Development and Implementation of ANN

The origin of artificial neural networks is based on learning technique that mimics the biological learning process occurring in the brain. Neural networks present a robust way to predict an actual value after a learning activity from a supplied sample set. The ANNs are based on a concept that combines a set of computational procedures with a theoretical basis in order to predict the unknown output parameter in various processes. Generally, neural networks are adapted to subordinate knowledge to observations or when data or activity is so complex that is not allows identifying an optimal solution in a reasonable time. It is difficult in each field of application and even for each task, to compare the use of neural networks versus other prediction techniques (e.g., statistical methods or a support vector machine) because, on the contrary to the conventional computational techniques, they are able to solve nonlinear and ill-defined problems. Many factors underlie this trend; most of them are related to reliability of the predictions, to the robustness and adaptability of the results as well as the learning ability of the neural process. In many cases the forecasts generated by ANNs, if correctly designed, significantly improve with increasing of dataset used as training subset. Consistently, in last years, the adoption of the ANNs in many business areas is increased exponentially and the number of publications, in high-level journals, was grown.

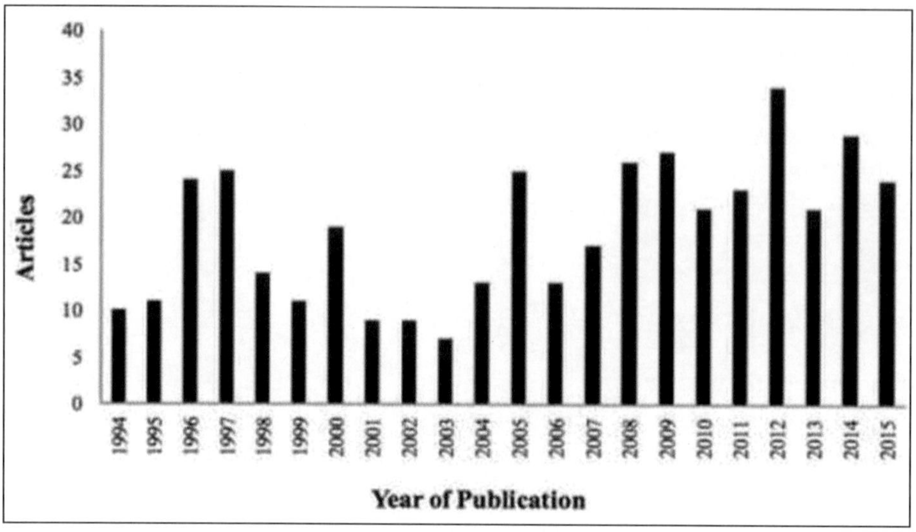

Distribution of ANN-papers by year.

Under engineering perspective, a 'good' ANN is based on models able to imitating the proprieties of natural systems, such as cognitive capabilities, flexibility, and robustness, ability to learn and fault tolerance. At this scope the structure and the behavior of the ANN required a study characterized by different hierarchical levels of organization as neurons, layers, synapses and cognition-behavior functions. Different areas of application are interested by the ANNs, some of them are astronomy, mathematics, physics chemistry, earth and space sciences, life and medical science and engineering. In recent years USA and EU countries, have approved different initiatives for the study of the human brain in these cases, the ANN, in various forms and at different levels, has been included in thesis research projects. The inter-disciplinary given by system adopted for dataset analysis and by the complexity computational required by the elaboration of the data, allows to design and simulate systems capable to satisfying the needs and the challenges of the real world. Japan in 2014 has been developed a project based named as Brain Mapping by Integrated Neurotechnologies for Disease Studies (Brain/MINDS), that will be integrated with new biomedical technologies and neural network systems. In Australia, a specific programme has also been set up with preliminary funds of around $250 million over 10 years with the goal of developing the world's first bionic brain (AusBrain) based on multilayer perceptron (MLP) system. There is also another ambitious initiative in China (Brainnetome), the goals of this are to simulate the brain networks for perception, memory, emotion and their disorders as well as to develop advanced technologies to achieve these goals.

Designing the ANN

An ANN is a computational model that establishes a relationship between process factors and output variables. Artificial neurons are combined through weights, which work as adjustable coefficients. There are many programs and frameworks, either of general purpose or that simulates functions or neural structures (e.g., IQR, NeuroSpaces, NNET, etc.) but there is not a specific simulator that is currently being used by the whole community since some different approaches are more suitable than others, on the basis of the research task being addressed. Moreover, most simulators can take full advantage of their computational capabilities on the basis of the features of the computer hardware to which it is installed.

This correlation depends by the fundamental features of the network, which define the way input and output are connected to each other. The network includes input layer, output layer and a certain number of hidden layers. The fundamental features of the network are:

- Dataset splitting, which identifies the subset data to be adopted for the training, the testing and the validation of the ANN development.
- Architecture, which determines the connections between layers and neurons.
- Learning algorithm, which determines the weights of the links between neurons.

Dataset Splitting

The appropriate data splitting can be handled as a statistical sampling problem. Therefore, various classical sampling techniques can be adopted in order to split the data in three subset for training, validation and testing of ANN, most commons are: Simple random sampling (SRS), Trial-and-error methods, Systematic sampling and Convenience sampling. The splitting strategy tries to overcome the high variance of the SRS by repeating the random sampling several times in order to minimize the mean square error (MSE) of the ANN. This technique is high time-consuming and requires significant computational costs. A subset (generally as big as 60% of the available experimental data composed by inputs/output pairs) is used for the ANN training. In this phase, the synaptic weights, which are the links between neurons, have a synaptic weight attached. They are updated repeatedly in order to reduce the error between the experimental outputs and the associated forecasts. A subset (generally as big as 20% of the available experimental data) is adopted for the ANN validation. In particular the validation sets allows to identifying the underlying trend of the training data subset. A subset (generally as big as 20% of the available experimental data) is adopted for testing the forecast reliability of the ANN in the learning phase. In order to deal with the overfitting problem that occurs when the network has memorized the training examples, but it has not learned to generalize to new situations, different approaches are suggested: reduce the number of hidden layers, improve the 'quality' of the training-subset adopted, introduce some noisy data into training set, etc. An efficient method is proposed for model establishment by means the identification of a low-dimension ANN learning matrix through the principal component analysis (PCA).

Network Architectures

An ANN is a computational model that establishes a relationship between process factors and output variables. Artificial neurons are combined through weights, which work as adjustable coefficients. This correlation depends by the fundamental features of the network, which define the way input and output are connected to each other. The network includes input layer, output layer and a certain number of hidden layers. The fundamental steps for the development of a network are:

- Dataset splitting, which identifies the subset data to be adopted for the training, the testing and the validation steps of the ANN development.
- Architecture, which determines the connections between layers and neurons.
- The learning algorithm, which determines the weights of the links between neurons.

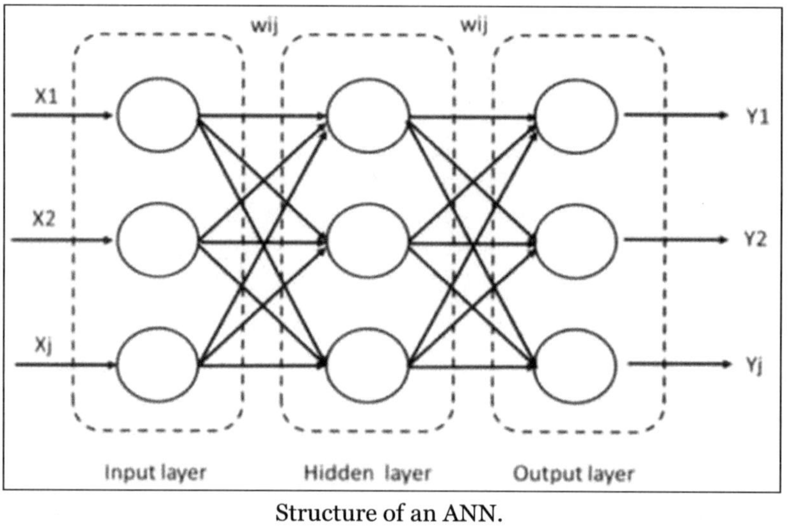

Structure of an ANN.

Based on the connection pattern (architecture), ANNs can be grouped into two categories:

- Feed-forward networks (e.g., single-layer perceptron, multilayer perceptron, radial basis function nets) in which there are not network connection as loops.
- Recurrent (or feedback) networks, in which different loops occur in network connections (e.g., competitive networks, Kohonen's SOM, Hopfield network, ART models).

The first one network are considered "static", in fact they produce only one set of output values rather than a sequence of values from a given input, and they worked in memory-less condition, this means that their response to an input is independent of the previous network state. The 'recurrent networks', on the other hand, are dynamic systems, in which the input pattern leads the network to enter in a new state, when a new input is introduced. Most popular network architecture in use today is the multilayer perceptron neural network (feed-forward network) where the output of a previous layer is the input to the next layer. In this case a biased sum of the weights assigned to different inputs, allows identifying the activation level that, through a transfer function, produces the corresponding output. The network thus has a simple interpretation as a form of input-output model, with the weights and thresholds (biases) the free parameters of the model. The design of the ANN architecture consists of identifying the kind of the structure (between feed-forward and recurrent architectures) and identifies the number of hidden layers and the number of neurons for each layer. On one hand, many neurons can lead to memorize the training sets with loss of the ANN's capability to generalize. On the other hand, a lack of neurons can inhibit the appropriate pattern classification. Many software allows to identify the best number of hidden layer and neurons (for each layer) through a 'trial-and-error' approach. In this case different architectures are iterative tested by software and for each of them, the software provide a "fitness bar" based on the inverse of the mean absolute error (MAE) computed on the testing set. In most cases the higher "fitness bar" identifies the best architecture.

Learning Algorithm

The purpose of the learning algorithm is to train the network to predict the output parameter given one or more input parameter. There are many types of neural network learning rules. There are three kind of learning algorithm, the first is known as supervised learning, in this case the algorithm allows predicting the output parameter on the basis of a set of known input-output pairs. Second algorithm is unsupervised learning, in this case the output is not given, the aim consisting of inferring a function in order to describe a hidden structure (e.g., clustering, anomaly detection, etc.). Therefore the output parameters are considered 'unlabelled' (the observations are not classified) and is not provided any evaluation about the prediction reliability ensured by the ANN. Third algorithm is named reinforcement learning, in this case a continue interaction between the learning system and the environment allows to identify the input-output mapping minimizing the performance scalar index. The approach is very similar to unsupervised learning (also in this case there are not given input-output pairs), reward or punishment signals are adopted for the prediction of output parameters. In most cases, the unsupervised learning allows to ensuring lower cost function. Three different methods, usually considered to be supervised learning methods, are described in this work: Quick Propagation (QP), Conjugate Gradient (CG) and Levenberg-Marquardt algorithm (LM).

QP is a heuristic modification of the standard back propagation, the output of the mth output node for the pth input pattern is given by opm.

$$O_{pm} = f\left(\sum_{k=1}^{K} \bar{\omega}_{km} o_{pk}\right)$$

where f is the activation sigmoidal function of equation $f(x)=\frac{1}{(1+e^{-x})}$, $\bar{\omega}_{km}$ is the weight between the mth output neuron and the kth hidden neuron. The value of opk depends by two parameters: the first is given by the weight between kth hidden neuron and the nth input neuron ($\bar{\omega}_{nk}$). The second parameter is xpn given by pth input pattern of nth neuron.

All network weights are updated after presenting each pattern from the learning data set.

As far as concern CG method, the learning algorithm starts with a random weight vector that is iteratively updated according the direction of the greatest rate of decrease of the error evaluated as ω (τ) in equation $\Delta\omega^{(\tau)} = -\eta \nabla E_{\omega}(r)$.

where E is the error function evaluated at $\omega^{(\tau)}$ and η is the arbitrary learning rate parameter. For each step (τ) the gradient is re-evaluated in order to reduce E. The performance of the gradient descent algorithm is very sensitive to the proper setting of the learning rate; in case η is too high the algorithm can oscillate and become unstable, for η too small the algorithm takes too long to converge. In this case an adaptive learning rate allows keeping the learning step size as large as possible, ensuring, in this way, the learning rate stable. The LM algorithm allows minimizing the squares of the differences (E) between the desirable output, identified as yd (t), and the predicted output yp (t). 'E' is given by the follow equation:

$$E = \frac{1}{2}\Sigma_{t=1}\left(y_p(t) - yd(t)\right)^2$$

LM algorithm is also adopted, which blends the 'Steepest Descent' method and the 'Gauss-Newton', therefore it can converge well even if the error surface is much more complex than the quadratic situation; ensuring, in many cases, speed and stability. LM algorithm can be presented as:

$$\omega k + 1 = \omega k - \left(J_k^T J_k + \mu I\right)^{-1} J_{kek}$$

where J is Jacobian matrix, μ is the 'combination coefficient' (always positive), I is the identity matrix and e represents the error vector. When μ is very small (nearly zero), Gauss-Newton algorithm is used. On the other hand, when μ is very large, steepest descent method is used.

Distributed Control System

Distributed Control System is a specially designed control system used to control complex, large and geographically distributed applications in industrial processes. In this, controllers are distributed throughout the entire plant area.

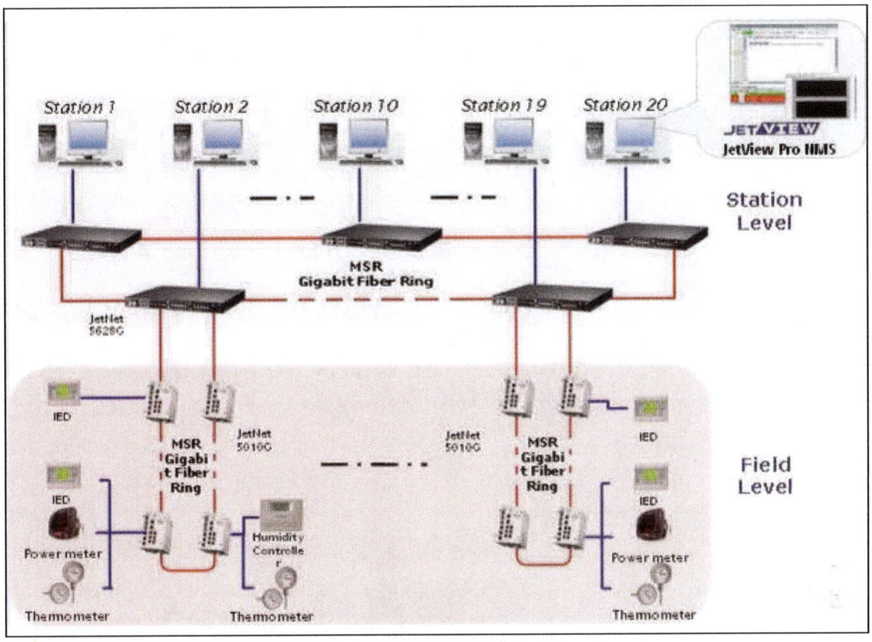

Distributed Control System.

These distributed controllers are connected to both field devices and operating PCs through high speed communication networks as shown in figure.

Discrete Field devices such as sensors and actuators are directly connected to input and output controller modules through communication bus. These field devices or smart instruments are capable of communicating with PLC's or other controllers while interacting with real world parameters like temperature, pressure, etc.

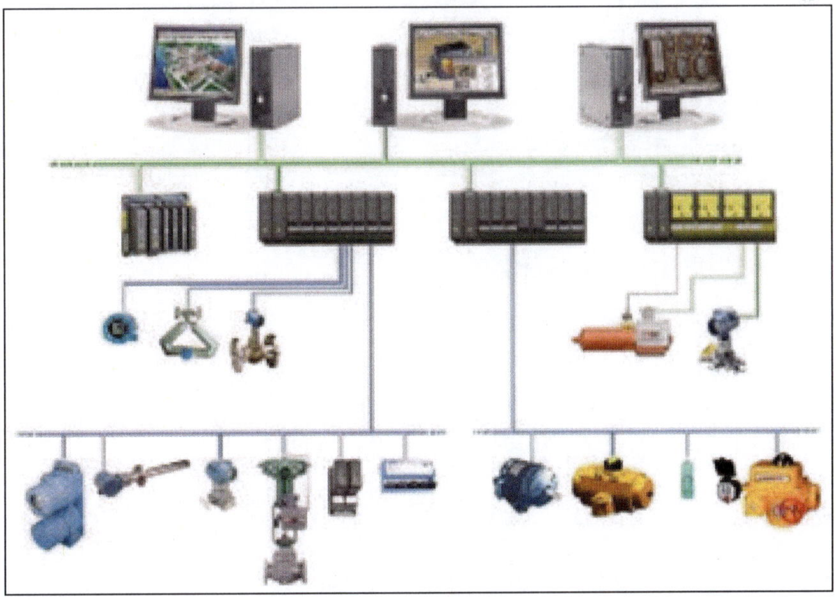

DCS Architecture.

Controllers are distributed geographically in various section of control area and are connected to operating and engineering stations which are used for data monitoring, data logging, alarming and controlling purpose via another high speed communication bus.

These communication protocols are of different types such as foundation filed bus, HART, Profibus, Modbus, etc. DCS provides information to multiple displays for user interface.

Four Basic Elements of Distributed Control System

Distributed Control System continuously interacts with the processes in process control applications ones it gets instruction from the operator. It also facilitates to variable set points and opening and closing of valves for manual control by the operator. Its human machine interface (HMI), face plates and trend display gives the effective monitoring of industrial processes.

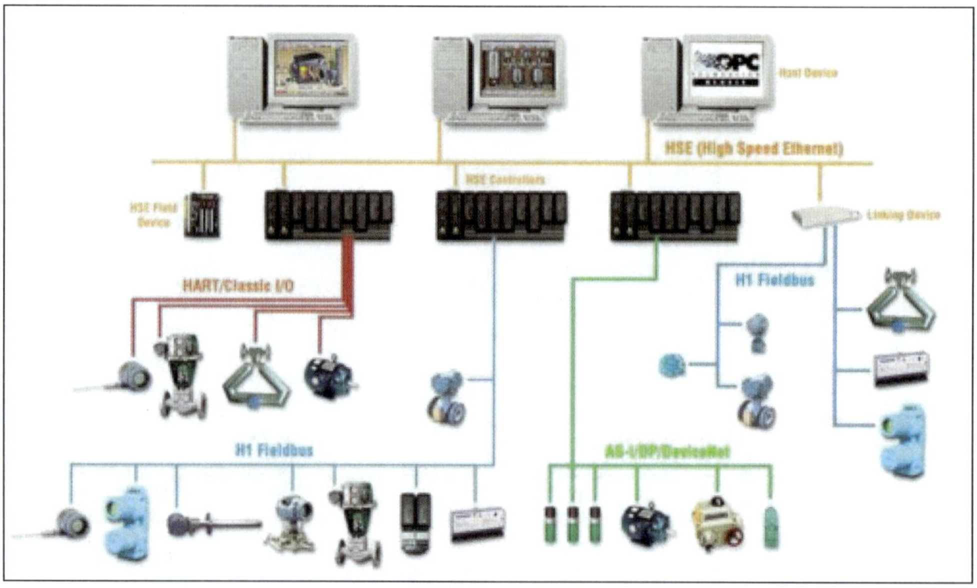

Elements of DCS.

Engineering PC or Controller

This controller is the supervisory controller over all the distributed processing controllers. Control algorithms and configuration of various devices are executed in this controller. Network communication between processing and engineering PC can be implemented by simplex or redundant configurations.

Distributed Controller or Local Control Unit

It can be placed near to field devices (sensors and actuators) or certain location where these field devices are connected via communication link. It receives the instructions from the engineering station like set point and other parameters and directly controls field devices.

It can sense and control both analog and digital inputs/outputs by analog and digital I/O modules. These modules are extendable according to the number of inputs and outputs. It collects the information from discrete field devices and sends this information to operating and engineering stations.

In above figure AC 700F and AC 800Fcontrollers acts as communication interface between field devices and engineering station. Most of the cases these act as local control for field instruments.

Operating Station or HMI

It is used to monitor entire plant parameters graphically and to log the data in plant database systems. Trend display of various process parameters provides the effective display and easy monitoring.

These operating stations are of different types such as some operating stations (PC's) used to monitor only parameters, some for only trend display, some for data logging and alarming requirements. These can also be configured to have control capabilities.

Communication Media and Protocol

Communication media consists of transmission cables to transmit the data such as coaxial cables, copper wires, fiber optic cables and sometimes it might be wireless. Communication protocols selected depends on the number of devices to be connected to this network.

For example, RS232 supports only for 2 devices and Profibus for 126 devices or nodes. Some of these protocols include Ethernet, DeviceNet, foundation filed bus, modbus, CAN, etc.

In DCS, two or more communication protocols are used in between two or more areas such as between field control devices and distributed controllers and other one between distributed controllers and supervisory control stations such as operating and engineering stations.

Architecture of Distributed Control System

As the name suggests, DCS has three main qualities. The first one is the distribution of various control functions into relatively small sets of subsystems, which are of semi-autonomous, and are interconnected through a high speed communication bus. Some of these functions include data acquisition, data presentation, process control, process supervision, reporting information, storing and retrieval of information.

The second attribute of DCS is the automation of manufacturing process by integrating advanced control strategies. And the third characteristic is the arranging the things as a system. DCS organizes the entire control structure as a single automation system where various subsystems are unified through a proper command structure and information flow.

These attributes of DCS can be observed in its architecture shown in the diagram below. The basic elements comprised in a DCS include engineering workstation, operating station or HMI, process control unit or local control unit, smart devices, and communication system.

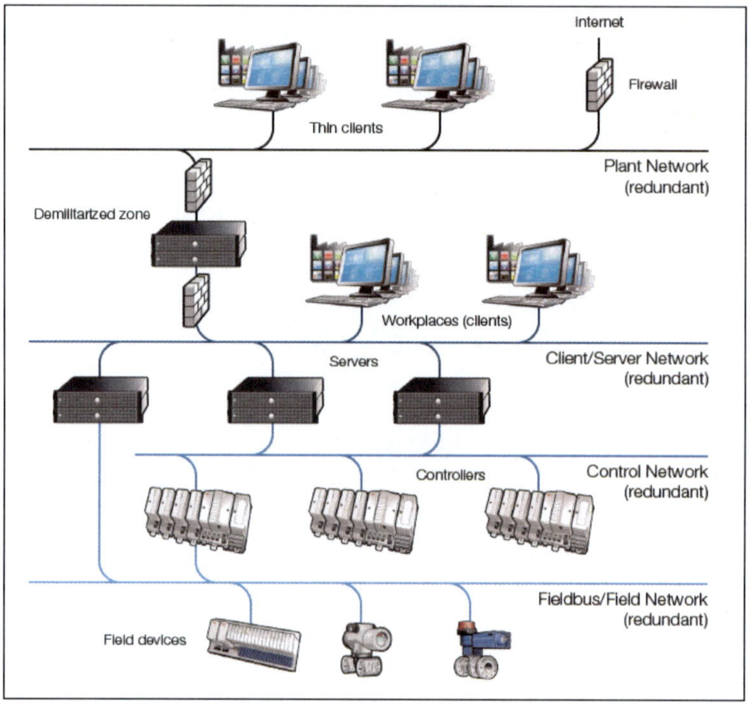

Engineering Workstation

It is the supervisory controller over the entire distributed control system. It can be a PC or any other computer that has dedicated engineering software (for example, control builder F engineering station in case of ABB freelance distributed control system).

This engineering station offers powerful configuration tools that allow the user to perform engineering functions such as creating new loops, creating various input and output points, modifying sequential and continuous control logic, configuring various distributed devices, preparing documentation for each input/output device, etc.

Operating Station or HMI

This is used to operate, monitor and control plant parameters. It can be a PC or any other monitoring device that has a separate software tool on which operator can view process parameter values and accordingly to take control action. For instance, it is a DigiVis software tool that can run on a simple PC-environment in case ABB DCS.

Operating stations can be a single unit or multiple units where a single unit performs functions like parameter value display, trend display, alarming, etc. while multiple units or PCs performs individual functions such as some PCs display parameters, some for trend archives, some for data logging and acquiring, etc.

Process Control Unit of DCS

It is also called as a local control unit, distribution controller, or process station. A distributed control system can consists of one or more process stations that can be extended with different types of I/O units. These controllers consist of a powerful CPU module, field bus or communication module with extended field bus capability and either direct or remote connected I/Os.

The field devices like sensors and actuators are connected to I/O modules of this unit. Some field devices can be directly connected to field bus (such as Profibus) without any I/O module, which can be termed as smart field devices.

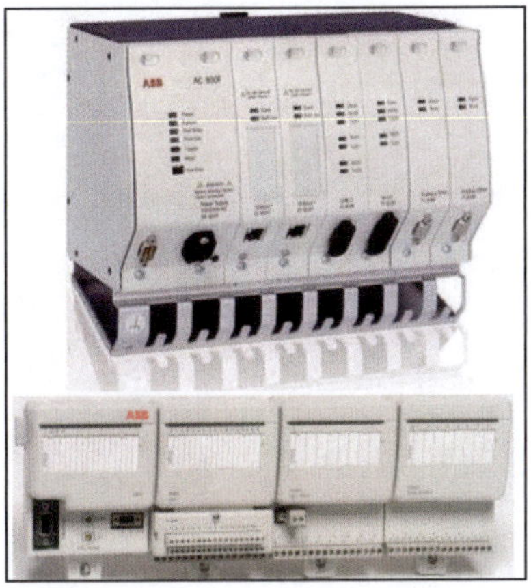

These units acquire the information from various sensors via input module, analyze and process it based on the control logic implemented and sends the output signals via output modules to have control on actuators and relays.

In case of <rel “nofollow” a href=”http://new.abb.com/control-systems/essential-automation/freelance/controller/ac800f” target=”_blank”>ABB DCS, AC800F module (consider, for instance) acts as process station, which is responsible for acquiring and controlling the data from the process. This unit consists of a power supply along with CPU section, Ethernet section, Profibus section and remote communication interface unit for I/Os interfacing as shown in the figure where first module is AC 800F unit and other one is remote I/O (also called as communication interface module).

Communication System

The communication medium plays a major role in the entire distributed control system. It interconnects the engineering station, operating station, process station and smart devices with one another. It carries the information from one station to another. The common communication protocols used in DCS include Ethernet, Profibus, Foundation Field Bus, DeviceNet, Modbus, etc.

It is not mandatory to use one protocol for entire DCS, some levels can use one network whereas some levels use different network. For instance, consider that field devices, distributed I/Os and process station are interconnected with Profibus while the communication among engineering station, HMI and process station carried though Ethernet as shown in the figure below.

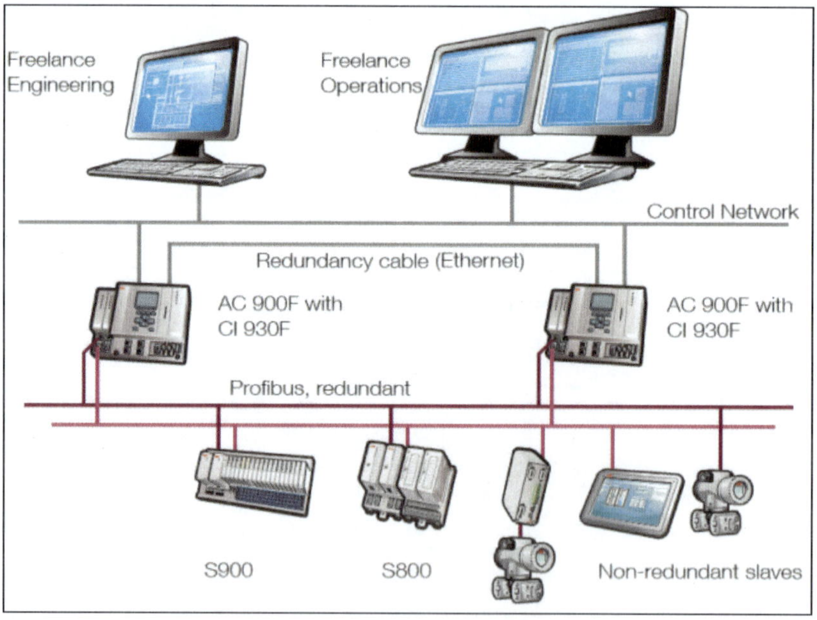

The major advantage of DCS is the redundancy of some or all levels of the control area. Most of the cases critical processes are installed with redundant controllers and redundant communication networks such that problem in main processing line should not affect the monitoring and control functions because of the redundant processing section.

Smart or Intelligent Devices

The intelligent field devices and field bus technology are advanced features of DCS technology that replaces traditional I/O subsystems (I/O modules). These smart devices embed the intelligence required for simple sensing and control techniques into the primary sensing and actuating devices. And hence it replaces the need for a DCS controller to perform routine sensing and control process.

These field devices can be directly connected to field bus so that sourcing of multiple measurements to the next higher level control station is possible via digital transmission line by eliminating extraneous hardware such as local I/O modules and controllers.

Working and Operation of DCS System

The operation of DCS goes like this; Sensors senses the process information and send it to the local I/O modules, to which actuators are also connected so as to control the process parameters. The information or data from these remote modules is gathered to the process control unit via field bus. If smart field devices are used, the sensed information directly transferred to process control unit via field bus.

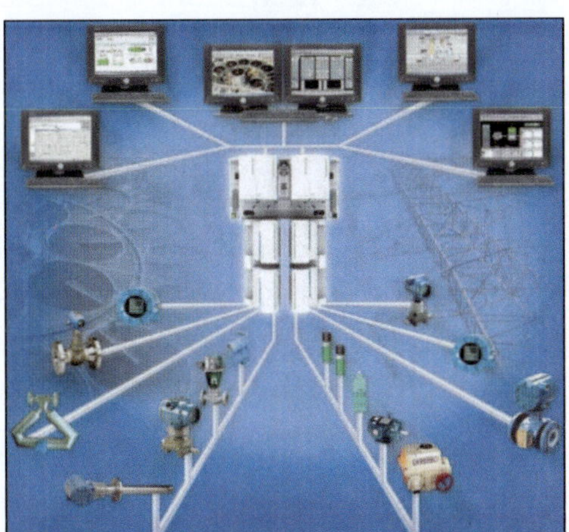

The collected information is further processed, analyzed and produces the output results based on the control logic implemented in the controller. The results or control actions are then carried to the actuator devices via field bus. The DCS configuring, commissioning and control logic implementation are carried at the engineering station as mentioned earlier. The operator able to view and send control actions manually at operation stations.

Difference between SCADA and DCS

Although both DCS and SCADA are monitoring and control mechanisms in industrial installations, they have different goals.

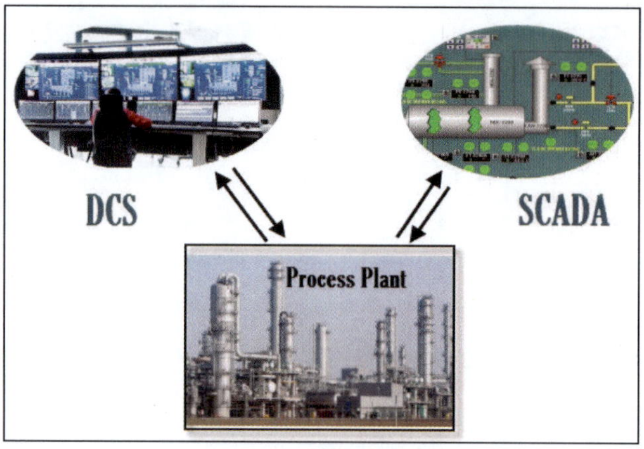

There exist some commonality between DCS and SCADA in terms of hardware and its components; however, there are certain requirements by the end applications that separate a robust and cost-effective DCS from the viable SCADA system. Some of the differences between DCS and SCADA are listed below:

- DCS is process oriented, whereas SCADA is data-gathering oriented. DCS emphasizes more on control of the process and it also consists of supervisory control level. And as a part of doing so, it presents the information to the operator. On the other hand, SCADA concentrates more on acquisition process data and presenting it to the operators and control centre.
- In DCS, data acquisition and control modules or controllers are usually located within a more confined area and the communication between various distributed control units carried via a local area network. SCADA generally covers larger geographical areas that use different communication systems which are generally less reliable than a local area network.
- DCS employs a closed loop control at process control station and at remote terminal units. But in case of SCADA there is no such closed loop control.
- DCS is process state driven where it scans the process in regular basis and displays the results to the operator, even on demand. On the other hand, SCADA is event driven where it does not scan the process sequentially, but it waits for

an event that causes process parameter to trigger certain actions. Hence, DCS does not keep a database of process parameter values as it always in connection with its data source, whereas SCADA maintains a database to log the parameter values which can be further retrieved for operator display and this makes the SCADA to present the last recorded values if the base station unable to get the new values from a remote location.

- In terms of applications, DCS is used for installations within a confined area, like a single plant or factory and for a complex control processes. Some of the application areas of DCS include chemical plants, power generating stations, pharmaceutical manufacturing, oil and gas industries, etc. On the other hand SCADA is used for much larger geographical locations such as water management systems, power transmission and distribution control, transport applications and small manufacturing and process industries.

In spite of these major differences, the modern DCS and SCADA systems come with common standard facilities while dealing process plant automation. However, the choice between DCS and SCADA depends on its client and end application requirement. But if the client choice between these two, by gaining equal requirement from the process, DCS is the economical choice as it help to reduce the cost and offer better control.

Important Features of DCS

To Handle Complex Processes

In factory automation structure, PLC-Programming Logic Controller is used to control and monitor the process parameters at high speed requirements. However due to limitation of number of I/O devices, PLC's cannot handle complex structure.

Handling Complex Processes.

Hence DCS is preferred for complex control applications with more number of I/O's with dedicated controllers. These are used in manufacturing processes where designing of multiple products are in multiple procedures such as batch process control.

System Redundancy

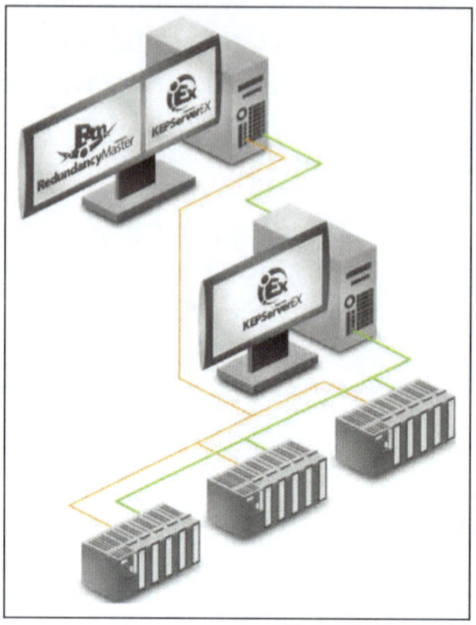

System Redundancy.

DCS facilitates system availability when needed by redundant feature at every level.

Resuming of the steady state operation after any outages, whether planned or unplanned is somewhat better compared to other automation control devices.

Redundancy raises the system reliability by maintaining system operation continuously even in some abnormalities while system is in operation.

Lot of Predefined Function Blocks

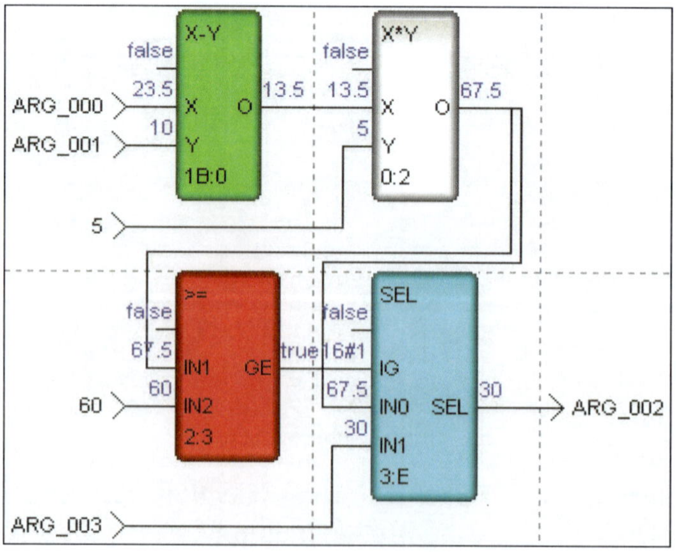

Predefined Function block.

DCS offers many algorithms, more standard application libraries, pre-tested and pre-defined functions to deal with large complex systems. This makes programming to control various applications being easy and consuming less time to program and control.

Powerful Programming Languages

It provides more number of programming languages like ladder, function block, sequential, etc for creating the custom programming based on user interest.

Similar to the SCADA system, DCS can also monitor and control through HMI's (Human Machine Interface) which provides sufficient data to the operator to charge over various processes and it acts as heart of the system. But this type of industrial control system covers large geographical areas whereas DCS covers confined area.

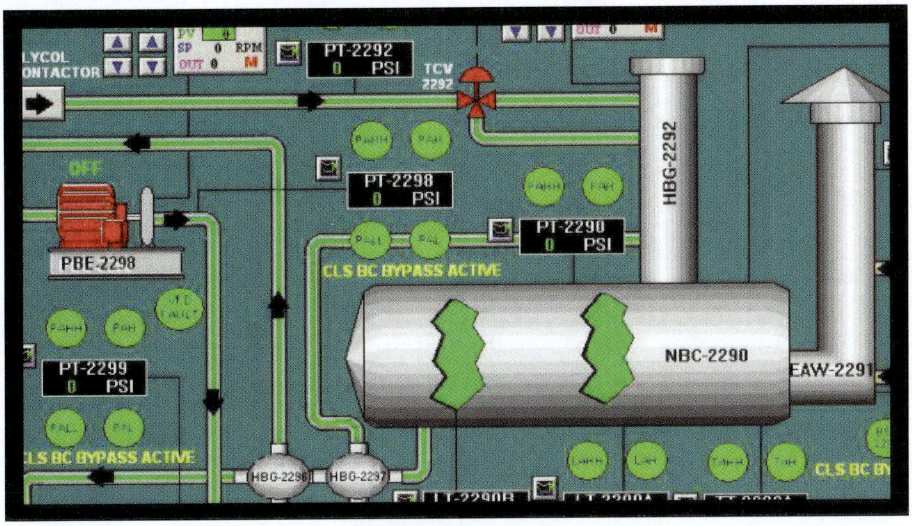

Sophisticated HMI.

DCS completely takes the entire process plant to control room as a PC window. Trending, logging and graphical representation of the HMI's give effective user interface. Powerful alarming system of DCS helps operators to respond more quickly to the plant conditions.

Scalable Platform

Structure of DCS can be scalable based on the number I/O's from small to large server system by adding more number of clients and servers in communication system and also by adding more I/O modules in distributed controllers.

System Security

Access to control various processes leads to plant safety. DCS design offers perfect secured system to handle system functions for better factory automation control. Security is also provided at different levels such as engineer level, entrepreneur level, operator level, etc.

Application of Distributed Control System

DCS system can be implemented in a simple application like load management using network of microcontrollers.

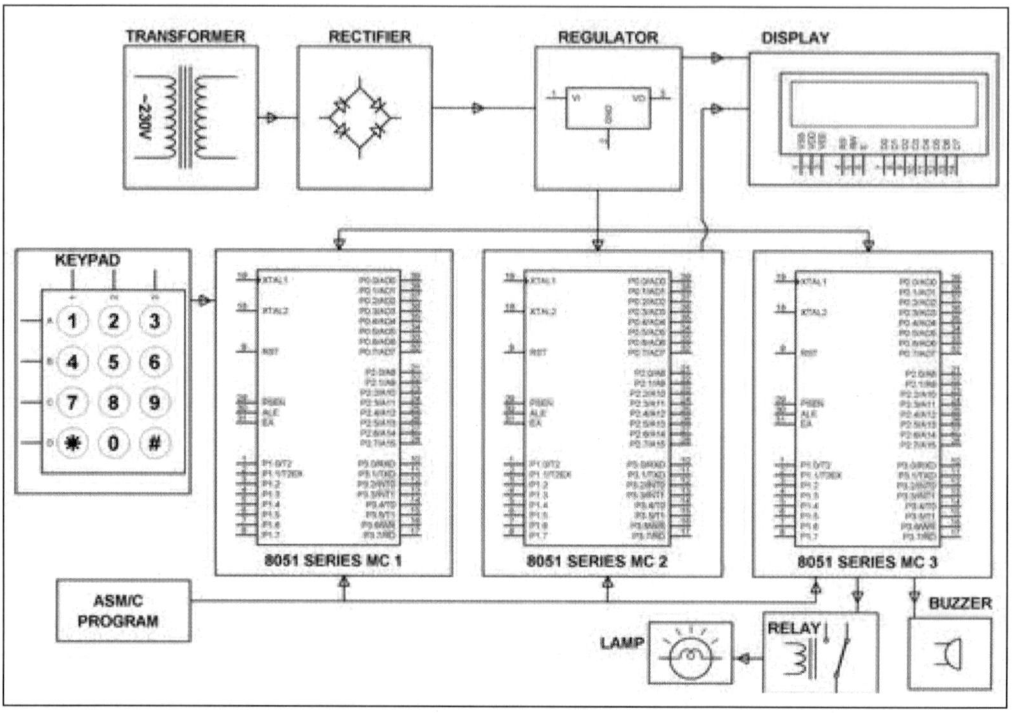

Application of DCS.

Here the input is given from a keypad to a microcontroller, which communicates with the other two microcontrollers. One of the microcontrollers is used to display the status of the process as well as the loads, while the other microcontroller controls the relay driver. The relay driver in turn drives the relay to operate the load.

References

- Industrial-control-system: trendmicro.com, Retrieved 04 May, 2019
- Ics-security, cyber-edu: forcepoint.com, Retrieved 25 April, 2019
- Discrete-control, engineering: sciencedirect.com, Retrieved 10 June, 2019
- Computer-Process-Control- 6389: brainkart.com, Retrieved 17 July, 2019
- Computer-numerical-control-CNC: searcherp.techtarget.com, Retrieved 09 February, 2019
- What-is-pid-controller-working-structure-applications: circuitdigest.com, Retrieved 05 March, 2019
- Scada-systems-work: elprocus.com, Retrieved 16 April, 2019
- What-is-plc: educba.com, Retrieved 16 August, 2019
- Distributed-control-system-dcs: electricaltechnology.org, Retrieved 18 April, 2019

4

Automation in Production Systems

Automation in production systems is concerned with operations and control of mechanical, industrial, electronic and computer based systems. Industrial automation uses information technologies and computers for handling operations in an industry. This chapter has been carefully written to provide an easy understanding of automation in production systems.

Some elements of the firm's production system are likely to be automated, whereas others will be operated manually or clerically. For our purposes here, automation can be defined as a technology concerned with the application of mechanical, electronic, and computer based systems to operate and control production.

The automated elements of the production system can be separated into two categories: (1) automation of the manufacturing systems in the factory and (2) computerization of the manufacturing support systems. In modern production systems, the two categories overlap to some extent, because the automated manufacturing systems operating on the factory floor are themselves often implemented by computer systems and connected to the computerized manufacturing support systems and management information system operating at the plant and enterprise levels. The term computer integrated manufacturing is used to indicate this extensive use of computers in production systems. The two categories of automation are shown in figure as an overlay on figure.

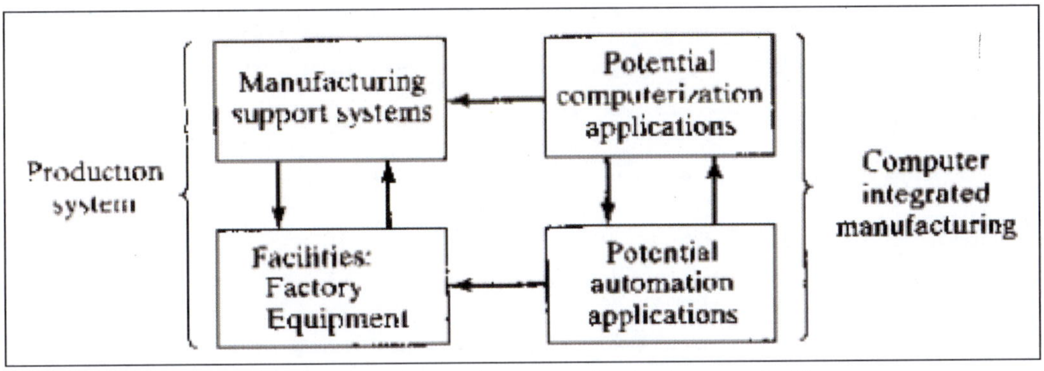

Opportunities of automation and computerization in a production system.

Automated Manufacturing Systems

Automated manufacturing systems operate in the factory on the physical product. They perform operations such as processing, assembly, inspection, or material handling, in some cases accomplishing more than one of these operations in the same system. They are called automated because they perform their operations with a reduced level of human participation compared with the corresponding manual process. In some highly automated systems, there is virtually no human participation. Examples of automated manufacturing systems include:

- Automated machine tools that process parts.
- Transfer lines that perform a series of machining operations.
- Automated assembly systems.
- Manufacturing systems that use industrial robots to perform processing or assembly operations.
- Automatic material handling and storage systems to integrate manufacturing operations.
- Automatic inspection systems for quality control.

Automated manufacturing systems can be classified into three basic types: (1) fixed automation, (2) programmable automation, and (3) flexible automation.

Fixed Automation

Fixed automation is a system in which the sequence of processing (or assembly) operations is fixed by the equipment configuration. Each of the operations in the sequence is usually simple, involving perhaps a plain linear or rotational motion or an uncomplicated combination of the two; for example, the feeding of a rotating spindle. It is the integration and coordination of many such operations into one piece of equipment that makes the system complex. Typical features of fixed automation are:

- High initial investment for custom engineered equipment.
- High production rates.
- Relatively inflexible in accommodating product variety.

The economic justification for fixed automation is found in products that are produced in very large quantities and at high production rates. The high initial cost of the equipment can be spread over a very large number of units, thus making the unit cost attractive compared with alternative methods of production. Examples of fixed automation include machining transfer lines and automated assembly machines.

Programmable Automation

In programmable automation, the production equipment is designed with the capability to change the sequence of operations to accommodate different product configurations.

The operation sequence is controlled by a program, which is a set of instructions coded so that they can be read and interpreted by the system. New programs can be prepared and entered into the equipment to produce new products. Some of the features that characterize programmable automation include:

- High investment in general purpose equipment.
- Lower production rates than fixed automation.
- Flexibility to deal with variations and changes in product configuration.
- Most suitable for batch production.

Programmable automated production systems are used in low and medium volume production. The parts or products are typically made in batches. To produce each new batch of a different product, the system must be reprogrammed with the set of machine instructions that correspond to the new product. The physical setup of the machine must also be changed: Tools must be loaded, fixtures must be attached to the machine table, and the required machine settings must be entered. This changeover procedure takes time. Consequently, the typical cycle for a given product includes a period during which the setup and reprogramming takes place, followed by a period in which the batch is produced. Examples of programmable automation include numerically controlled (NC) machine tools, industrial robots, and programmable logic controllers.

Flexible Automation

Flexible automation is an extension of programmable automation. A flexible automated system is capable of producing a variety of parts (or products) with virtually no time lost for changeovers from one part style to the next. There is no lost production time while reprogramming the system and altering the physical setup (tooling, fixtures, machine settings). Consequently, the system can produce various combinations and schedules of parts or products instead of requiring that they be made in batches. What makes flexible automation possible is that the differences between parts processed by the system are not significant. It is a case of soft variety, so that the amount of changeover required between styles is minimal. The features of flexible automation can be summarized as follows:

- High investment for a custom engineered system.
- Continuous production of variable mixtures of products.
- Medium production rates.
- Flexibility to deal with product design variations.

Examples of flexible automation are the flexible manufacturing systems for performing machining operations that date back to the late 1960s.

The relative positions of the three types of automation for different production volumes and product varieties are depicted in figure. For low production quantities and new product introductions, manual production is competitive with programmable automation.

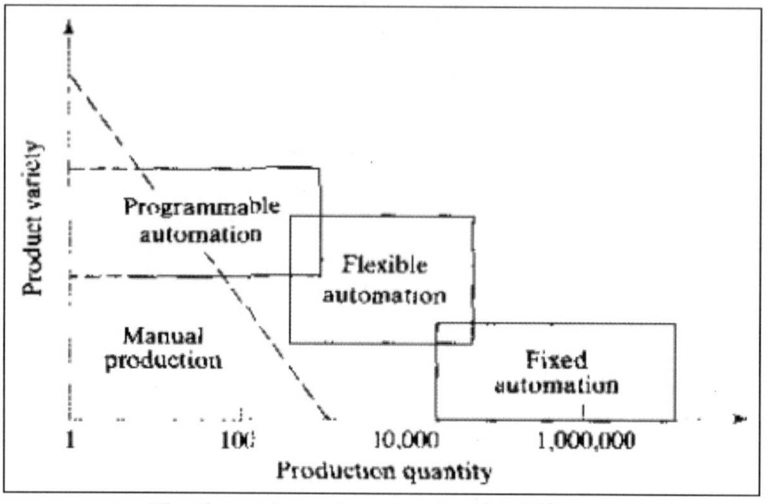

The three types of automation for different production volumes and product variety.

Computerized Manufacturing Support Systems

Automation of the manufacturing support systems is aimed at reducing the amount of manual and clerical effort in product design, manufacturing planning and control, and the business functions of the firm. Nearly all modern manufacturing support systems are implemented using computer systems. Indeed, computer technology is used to implement automation of the manufacturing systems in the factory as well. The term computer— integrated manufacturing (CIM) denotes the pervasive use of computer systems to design the products, plan the production, control the operations, and perform the various business—related functions needed in a manufacturing firm. True CIM involves integrating all of these functions in one system that operates throughout the enterprise. Other terms are used to identify specific elements of the CIM system. For example, computer aided design (CAD) denotes the use of computer systems to support the product design function. Computer aided manufacturing (CAM) denotes the use of computer systems to perform functions related to manufacturing engineering, such as process planning and numerical control part programming. Some computer systems perform both CAD and CAM, and so the term CAD/CAM is used to indicate the integration of the two into one system. Computer—integrated manufacturing includes CAD/CAM, but it also includes the firm's business functions that are related to manufacturing.

Let us attempt to define the relationship between automation and CIM by developing a conceptual model of manufacturing. In a manufacturing firm, the physical production activities that take place in the factory can be distinguished from the information—processing activities, such as product design and production planning, that usually occur in an office environment. The physical activities include all of the processing, assembly, material handling, and inspection operations that are performed on the product in the factory. These operations come in direct contact with the product during manufacture. The relationship between the physical activities and the information—processing activities in our model is depicted in figure. Raw materials flow into one end of the factory and finished products flow out the other end. The physical activities take place inside the factory. In our model, the information—processing activities form a ring that surrounds the factory, providing the data and knowledge required to successfully produce the product.

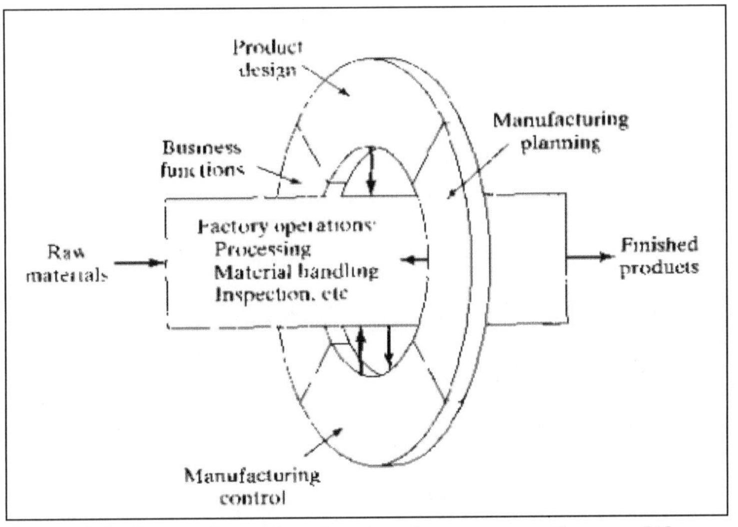

Model of manufacturing showing factory operations and the information-processing activities for manufacturing support.

These information—processing activities are accomplished to implement the four basic manufacturing support functions identified earlier: (1) business functions, (2) product design, (3) manufacturing planning, and (4) manufacturing control. These four functions form a cycle of events that must accompany the physical production activities but do not directly touch the product.

Reasons for Automating

Companies undertake projects in manufacturing automation and computer integrated manufacturing for a variety of good reasons. Some of the reasons used to justify automation are the following:

- To increase labor productivity: Automating a manufacturing operation usually increases production rate and labor productivity. This means greater output per hour of labor input.
- To reduce labor cost: Ever increasing labor cost has been and continues to be the trend in the world's industrialized societies. Consequently, higher investment

in automation has become economically justifiable to replace manual operations. Machines are increasingly being substituted for human labor to reduce unit product cost.

- To mitigate the effects of labor shortages: There is a general shortage of labor in many advanced nations, and this has stimulated the development of automated operations as a substitute for labor.
- To reduce or eliminate routine manual and clerical tasks: An argument can be put forth that there is social value in automating operations that are routine, boring, fatiguing, and possibly irksome. Automating such tasks serves a purpose of improving the general level of working conditions.
- To improve worker safety: By automating a given operation and transferring the worker from active participation in the process to a supervisory role, the work is made safer. The safety and physical wellbeing of the worker has become a national objective with the enactment of the Occupational Safety and Health Act (OSHA) in 1970. This has provided an impetus for automation.
- To improve product quality: Automation not only results in higher production rates than manual operations; it also performs the manufacturing process with greater uniformity and conformity to quality specifications. Reduction of fraction defect rate is one of the chief benefits of automation.
- To reduce manufacturing lead time: Automation helps to reduce the elapsed time between customer order and product delivery, providing a competitive advantage to the manufacturer for future orders. By reducing manufacturing lead time, the manufacturer also reduces workin process inventory.
- To accomplish processes that cannot be done manually: Certain operations cannot be accomplished without the aid of a machine. These processes have requirements for precision, miniaturization, or complexity of geometry that cannot be achieved manually. Examples include certain integrated circuit fabrication operations, rapid prototyping processes based on computer graphics (CAD) models, and the machining of complex, mathematically defined surfaces using computer numerical control. These processes can only be realized by computer controlled systems.
- To avoid the high cost of not automating: There is a significant competitive advantage gained in automating a manufacturing plant. The advantage cannot easily be demonstrated on a company's project authorization form. The benefits of automation often show up in unexpected and intangible ways, such as in improved quality, higher sales, better labor relations, and better company image. Companies that do not automate are likely to find themselves at a competitive disadvantage with their customers, their employees, and the general public.

Industrial Automation

Industrial Automation is a set of industrial processes that run automatically without human intervention. It is achieved by various means like pneumatic, electronic, hydraulic, computer and electrical devices. Some advantages of Industrial Automation are:

- It saves labor cost.
- It increases productivity and throughput of the process.
- It enhances consistency of work.
- It improves robustness.

These are generally done when a high degree of accuracy is required and when the tasks that are done in a dangerous environment.

Nowadays, Industrial Automation is common in industries to produce high-quality goods at economical prices. The tasks are classified into the following categories, material handling, processing, assembly of materials, and inspection. The system can be programmed in several ways thus it is a customized way to fulfill the needs and requirements of the clients. Although the automation industry enjoys remarkable profits, it has some disadvantages too. For instance, the setup is costly, maintenance is difficult and experienced operators are required to handle the operations. However, the cost of the process may depend on the technology or the engine that is automated.

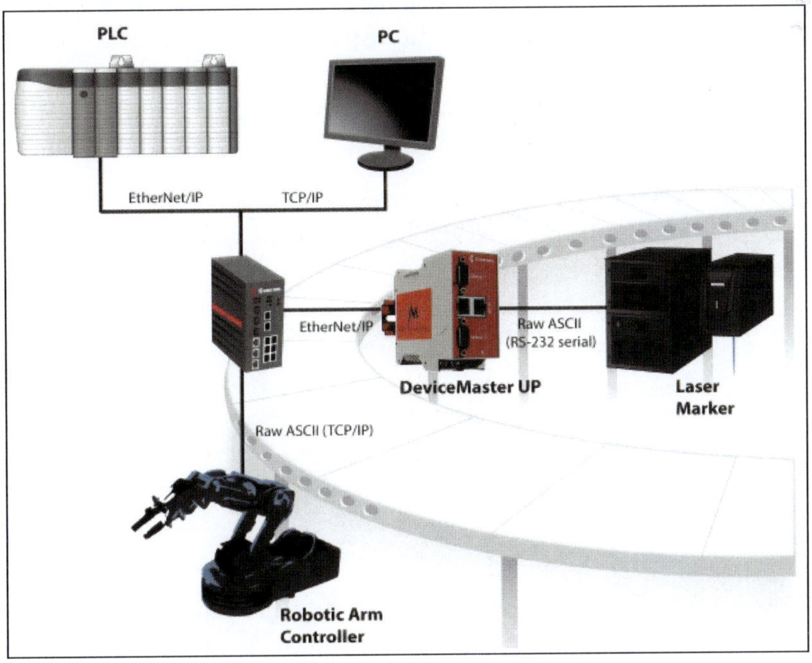

Advantages of Automation System

Increase Productivity

Automation of factory or manufacturing or process plant improves production rate through a better control of production. It helps to produce mass production by drastically reducing assembly time per product with a greater production quality. Therefore, for a given labor input it produces a large amount of output.

Provide Optimum Cost of Operation

Integration of various processes in industry with automated machineries, minimizes cycle times and effort and hence the need of human labor gets reduced. Thus the investment on employees has been saved with automation.

Improve Product Quality

Since the automation reduces the human involvement, the possibility of human errors also gets eliminated. Uniformity and product quality with a greater conformity can be maintained with automation by adaptively controlling and monitoring the industrial processes in all stages right from inception of a product to an end product.

Reduce Routine Checks

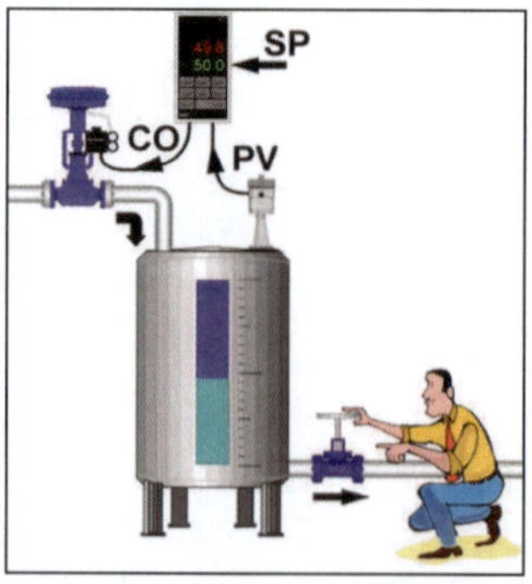

Automation completely reduces the need for manual checking of various process parameters. By taking advantage of automation technologies, industrial processes automatically adjusts process variables to set or desired values using closed loop control techniques.

Raise the Level of Safety

Industrial automation increases the level of safety to personnel by substituting them with automated machines in hazardous working conditions. Traditionally, industrial robots and robotic devices are implemented in such risky and hazardous places.

Hierarchy of an Industrial Automation System

Industrial automation systems can be very complex in nature, having large number of devices working in synchronization with automation technologies. The figure below describes the hierarchical arrangement of the automation system consisting of different hierarchical levels.

Field Level

It is the lowest level of the automation hierarchy which includes the field devices like sensors and actuators. The main task of these field devices is to transfer the data of processes and machines to the next higher level for monitoring and analysis. And also it includes the controlling of process parameter through actuators. For instance, we can describe this level as eyes and arms of a particular process.

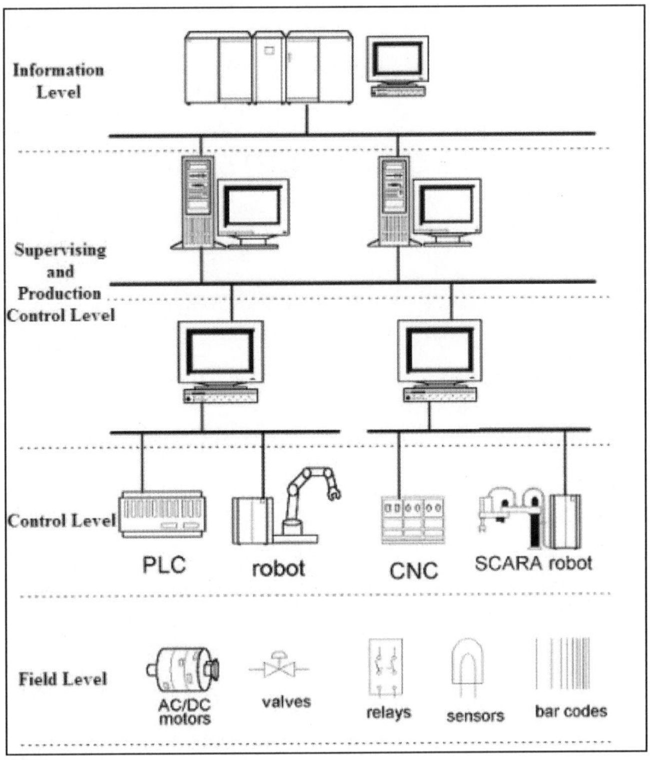

Sensors convert the real time parameters like temperature, pressure, flow, level, etc. into electrical signals. This sensor data further transferred to the controller so as to monitor and analyze the real time parameters. Some of the sensors include thermocouple, proximity sensors, RTDs, flow meters, etc.

On other hand actuators converts the electrical signals (from the controllers) into mechanical means to control the processes. Flow control valves, solenoid valves, pneumatic actuators, relays, DC motors and servo motors are the examples of actuators.

Control Level

This level consists of various automation devices like CNC machines, PLCs, etc., which acquires the process parameters from various sensors. The automatic controllers drive the actuators based on the processed sensor signals and program or control technique.

Programmable Logic Controllers (PLCs) are most widely used robust industrial controllers which are capable of delivering automatic control functions based on input from sensors. It consists of various modules like CPU, analog I/O, digital I/O and communication modules. It allows the operator to program a control function or strategy to perform certain automatic operation on process.

Supervising and Production Control Level

In this level, automatic devices and monitoring system facilitates the controlling and intervening functions like Human Machine Interface (HMI), supervising various parameters, setting production targets, historical archiving, setting machine start and shutdown, etc.

Mostly, either Distribution Control System (DCS) or Supervisory Control and Data Acquisition (SCADA) HMIs are popularly used in this level.

Information or Enterprise Level

This is the top level of the industrial automation which manages the whole automation system. The tasks of this level include production planning, customer and market analysis, orders and sales, etc. So it deals more with commercial activities and less with technical aspects.

And also industrial communication networks are most prominent in industrial automation systems which transfer the information from one level to the other. So these are present in all the levels of automation system to provide continuous flow of information. This communication network can be different one level to the other. Some of these networks include RS485, CAN, DeviceNet, Foundation Field bus, Profibus, etc.

From the above hierarchy we can conclude that there is continuous information flow from high level to low level and vice-versa. If we assume this graphical way, it is like a pyramid in which as we go up, the information gets aggregated and while going down, we get detailed information about the process.

Types of Industrial Automation

Industrial automation is the use of computer and machinery aided systems to operate the various industrial operations in a well-controlled manner. Depends on the operations involved, the industrial automation systems are majorly classified into two types, namely process plant automation and manufacturing automation.

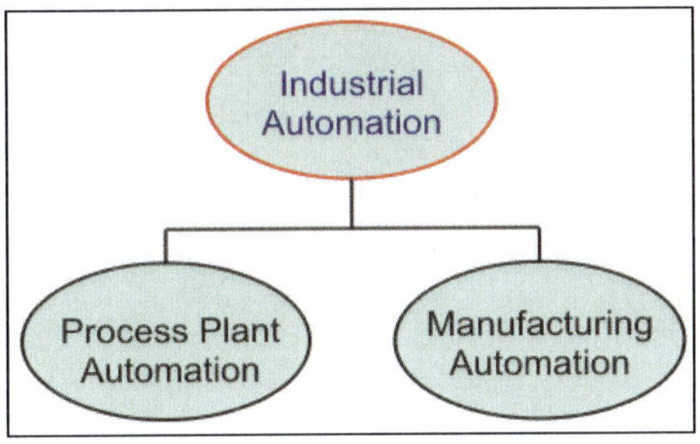

Process Plant Automation

In process industries, the product results from many chemical processes based on some raw materials. Some of the industries are pharmaceuticals, petrochemical, cement industry, paper industry, etc. Thus the overall process plant is automated to produce the high quality, more productive, high reliable control of the physical process variables.

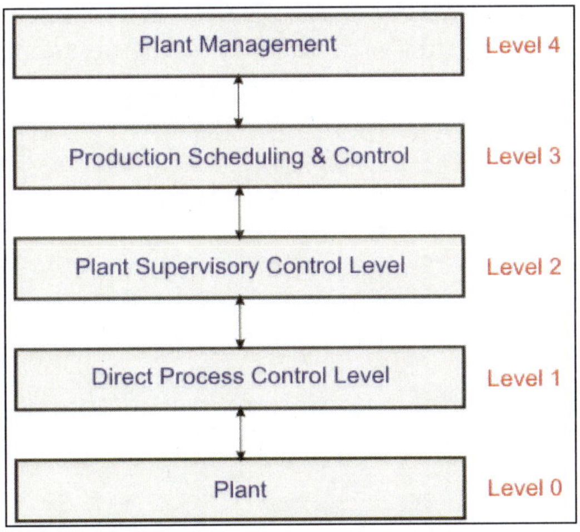

Process Plant Automation hierarchy.

The above figure shows the process automation system hierarchy. It consists of various layers representing widespread components in a process plant.

- Level 0 or Plant: This level consists of machines which are closest to processes. In this, sensors and actuators are used to translate the signals from the machines and physical variables for the purpose of analysis and to produce the control signals.

- Direct Process Control: In this level, automatic controllers and monitoring systems acquire the process information from sensors and correspondingly drive the actuator systems. Some of the tasks of this level are:
    - Data acquisition,
    - Plant monitoring,
    - Data checking,
    - Open and closed loop control,
    - Reporting.
- Plant Supervisory Control: This level commands the automatic controllers by setting the targets or set points. It looks after the control equipment for optimal process control. Some of the tasks of this level are:
    - Plant monitoring performance,
    - Optimal process control,
    - Plant coordination,
    - Failure detection, etc.
- Production Scheduling and Control: This level solves the decision-making problems like resource allocation, production target, maintenance management, and so on. Tasks of this level include:
    - Production dispatch,
    - Inventory control,
    - Production supervision, production reporting, etc.
- Plant Management: This is the higher level of the process plant automation. It deals more with commercial activities than technical activities. Tasks of this level include:
    - Market and Customer analysis,
    - Orders and sale statistics,
    - Production planning,
    - Capacity and order balance, etc.

Manufacturing Automation System

The manufacturing industries make the product out of materials using machines/robotics. Some of these manufacturing industries include textile and clothing, glass and ceramic, food and beverages, paper making, etc. New trends in manufacturing systems have been using automation systems at every stage such as material handling, machining, assembling, inspection, and packaging. With the computer-aided control and industrial robotic systems, manufacturing automation becomes very flexible and efficient.

The figure below shows the manufacturing automation system hierarchy in which all functional levels are automated by using different automation tools.

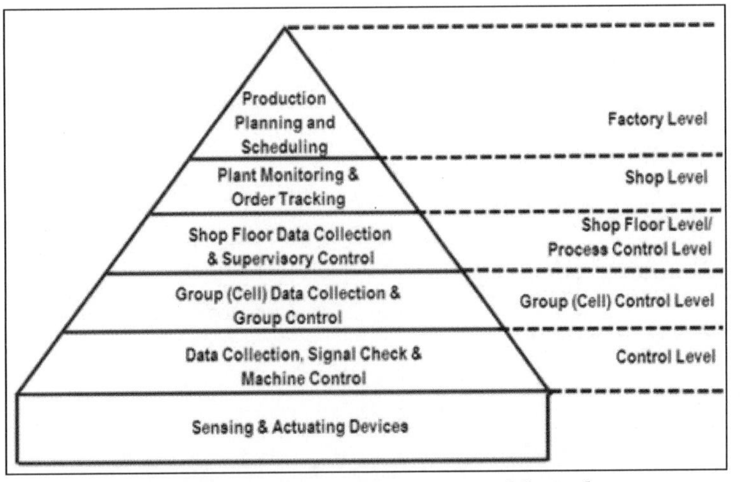

Manufacturing automation systems hierarchy.

Below is an explanation of each level in the manufacturing automation system hierarchy:

- Machinery Level: In this level various sensing and actuating devices controls the manufacturing process. It is an instrumentation level of machine control. Tasks of this level include data collection, signal check, and machine control.
- Cell or Group Level: This is another automation level at which operation of a group of machines within manufacturing cells are co-ordinated. Various automated controllers like PLCs are employed for such control of machines.
- Shop Floor Level: It is a supervisory automated level where supervision and coordination of several manufacturing cells are carried out.
- Plant Level: This automation level performs the activities of production monitoring, control, and scheduling, etc. HMIs employed at this level facilitate to control all the manufacturing process variables remotely.
- Enterprise Level: This level does all the management related activities such as production planning and scheduling, etc.

Industrial Automation Equipment

Industrial automation (IA) is an integrated, flexible and low-cost system platform that consists of various equipment and elements which perform a wide variety of functions like sensing, control, supervision and monitoring related to industrial processes. The figure below shows the structure of industrial automation which describes the various functional elements of IA.

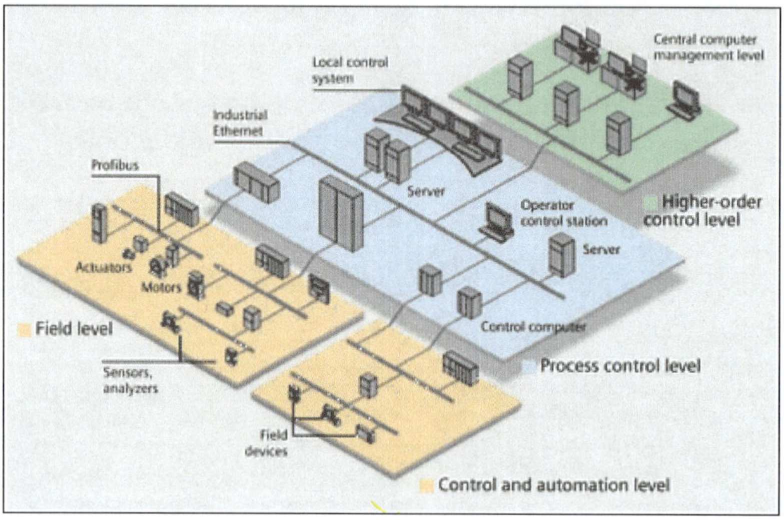

Sensing and Actuating Elements

The sensors or sensing elements convert the physical process variables such as flow, pressure, temperature, etc. into electrical or pneumatic form. Various sensors include thermocouples, Resistor Temperature Detectors (RTDs), strain gauges, etc. The signals from these sensors are used for processing, analyzing, and decisions in order to produce the control output. The various control techniques are implemented to produce the required output by comparing the current sensed process variable with set values. Finally, the controllers produce the computed outputs and are applied as electrical or pneumatic signal inputs to the actuating elements. Actuators convert the electrical or pneumatic signals to the physical process variables. Some of the actuators include control valves, relays, motors, etc.

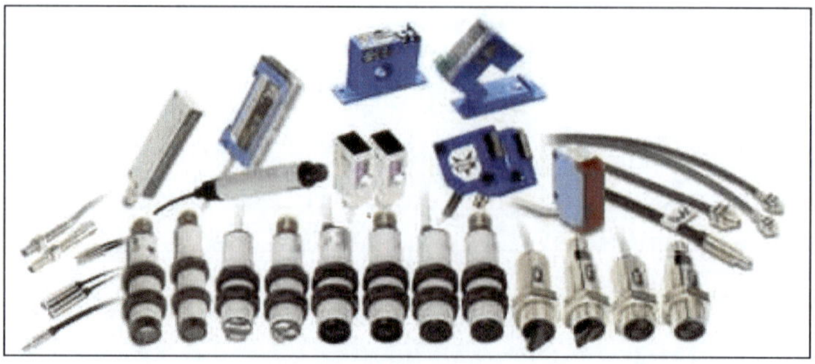

A special category of the instruments is smart instruments which are integrated systems of sensing or actuating elements with the capability of communicating with field buses. These smart devices consist of a signal conditioning circuit internally and facilitate connecting directly to the communication link in the industrial bus system.

Control System Elements

These are the microprocessor-based electronic controllers or simply industrial computers that accept the signals from various sensors as well as command signals from supervisory systems or from human operators. These controllers can be continuous control systems or sequential/logic control depends on the structure of control nature. The controller processes the sensing values and supervisory values and depends on the control structure, it produce the control output to various actuating devices.

A modern type of control device used in automation systems is the Programmable Logic Controller (PLC). PLCs come with dedicated software so that these are capable of being programmed to perform corresponding control operations. PLCs have rugged CPU, digital I/O, analog I/O and communication modules such that they can operate at industry environment conditions to control the various process parameters.

Human Machine Interface or operator interface is a graphical interface for the operators which display the process information such as process variable status, logging results to the database, generating alarm signals, etc. SCADA is one of the graphical user interfaces which remotely control the industrial operations. Also, Distributed control systems (DCS) provide their own HMI for the graphical display of various industrial parameters.

Supervisory Control Elements

Supervisory control performs higher level control over the automatic controllers which further controls the smaller subsystems. The major elements of this level are process station PCs, and Human Machine Interfaces. These process station PCs are responsible for functions like set point computations, performance monitoring, diagnostics, start-up, shutdown, and other emergency operations.

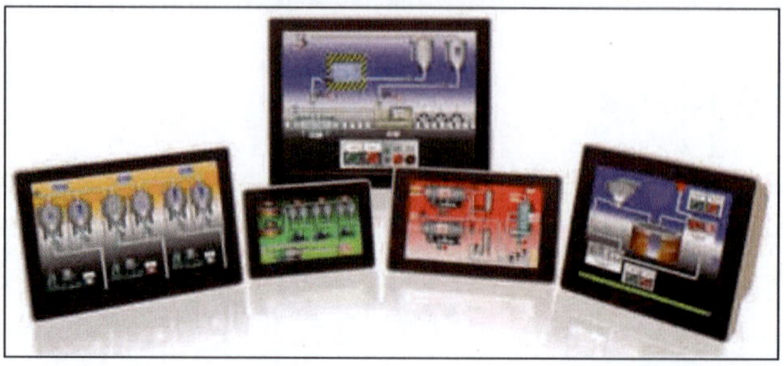

Human Machine Interface or operator interface is a graphical interface for the operators which display the process information such as process variable status, logging results to the database, generating alarm signals, etc. SCADA is one of the graphical user interfaces which remotely controls the industrial operations. Also, Distributed control systems (DCS) provide their own HMI for the graphical display of various industrial parameters.

Advantages of Industrial Automation

Manufacturers face many challenges in today's globally competitive business landscape. Some of these challenges include harsh manufacturing environments (in a world which is increasingly focused on safety – and rightly so), increasingly complex supply chains, meeting the latest energy efficient standards, and competing with companies with very small marginal costs.

Many of these reasons drive manufacturers towards industrial automation. The advantages to Industrial Automation include:

- Increased labor productivity,
- Improved product quality,
- Reduced labor or production cost,
- Reduced routine manual tasks,
- Improved safety,
- Assisted remove monitoring.

Increased Labor Productivity

Automation increases the production rate by producing greater output for a given labor input. It is not possible for human workers to work for long hours without losing accuracy. On the other hand, without compromising on accuracy, automated control systems able to work for long hours. Hence, increased productivity and efficiency per hour of labor input.

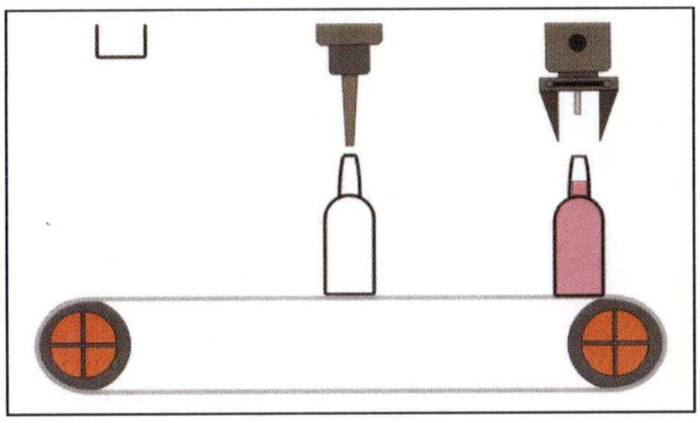

Improved Product Quality

One of the chief benefits of the automation is that the reduction of fraction defect rate. With the manual operation of manufacturing process, there may be a compromise on quality specifications of the product. But the automation system performs operations with greater conformity and uniformity to the quality specifications. By using the automation systems, industrial processes are controlled and monitored at all stages in order to produce a qualitative end product.

Reduced Labor or Production Cost

The automated systems help the industries to save a great deal in the long term by substituting automated machinery in place of human labor so that unit production cost is reduced. Automation equipment running smoothly or uniformly 24 × 7 not only increases the productivity, but also consequently results in an excellent return on investment by saving salaries, workforce costs, pensions and costs with employees. The automated system also reduces the labor shortage by substituting automated operations in place of labor.

Reduced Routine Manual Tasks

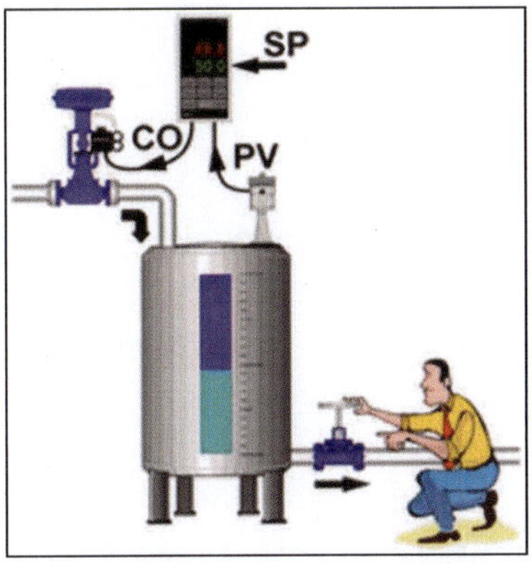

In many industrial applications, process variables like temperature, liquid level, pressure, etc. are to be periodically monitored as a routine task to maintain their set levels.

Thus an automation system creates the automatic working condition by employing closed-loop control systems.

Improved Safety

By implementing an automated system, work is made safer by transferring the worker from an active participation location in the process to the supervising role. The automated machines are able to work in hazardous environments and other extreme environments. Also, these systems make use of industrial robots in place of human workers, especially in life-threatening conditions (chemical and high-temperature conditions). Thus an industrial automation system prevents the accidents and injuries to the workers.

Assist Remote Monitoring

Most of the industrial operations have to be controlled remotely for convenient and long distance monitor and control of process variables. For such cases, automated systems provide a communication link between the process area and supervising (monitor and control) area, thereby allowing operators to control and monitor the industrial processes from a remote location. The best example of this remote control is the automated electric power grid control.

References

- Automation-in-Production-Systems- 6373: brainkart.com, Retrieved 16 January, 2019
- Industrial-automation- 8906: petropedia.com, Retrieved 15 August, 2019
- What-is-industrial-automation: electricaltechnology.org, Retrieved 07 May, 2019
- Industrial-automation: electrical4u.com, Retrieved 10 July, 2019

5

Applications of Automation

There are wide-range of applications of automation in the fields of retail, mining, highway systems, car and home systems, waste management, power production, etc. This chapter closely examines the diverse applications of automation to provide an extensive understanding of the subject.

Automation in Daily Life

Communications

One of the earliest practical applications of automation was in telephone switching. The first switching machines, invented near the end of the 19th century, were simple mechanical switches that were remotely controlled by the telephone user pushing buttons or turning a dial on the phone. Modern electronic telephone switching systems are based on highly sophisticated digital computers that perform functions such as monitoring thousands of telephone lines, determining which lines require service, storing the digits of each telephone number as it is being dialed, setting up the required connections, sending electrical signals to ring the receiver's phone, monitoring the call during its progress, and disconnecting the phone when the call is completed. These systems also are used to time and bill toll calls and to transmit billing information and other data relative to the business operations of the phone company. In addition to the various functions mentioned, the newest electronic systems automatically transfer calls to alternate numbers, call back the user when busy lines become free, and perform other customer services in response to dialed codes. These systems also perform function tests on their own operations, diagnose problems when they arise, and print out detailed instructions for repairs.

Other applications of automation in communications systems include local area networks, communications satellites, and automated mail-sorting machines. A local area network (LAN) operates like an automated telephone company within a single building or group of buildings. Local area networks are generally capable of transmitting not only voice but also digital data between terminals in the system. Communications satellites have become essential for communicating telephone or video signals across great distances. Such communications would not be possible without the automated guidance systems that place and retain the satellites in predetermined orbits. Automatic mail-sorting machines have been developed for use in many post offices throughout the world to read codes on envelopes and sort the envelopes according to destination.

Transportation

Automation has been applied in various ways in the transportation industries. Applications include airline reservation systems, automatic pilots in aircraft and locomotives, and urban mass-transit systems. The airlines use computerized reservation systems to continuously monitor the status of all flights. With these systems, ticket agents at widely dispersed locations can obtain information about the availability of seats on any flight in a matter of seconds. The reservation systems compare requests for space with the status of each flight, grant space when available, and automatically update the reservation status files. Passengers can even receive their seat assignments well in advance of flight departures.

Nearly all commercial aircraft are equipped with instruments called automatic pilots. Under normal flying conditions, these systems guide an airplane over a predetermined route by detecting changes in the aircraft's orientation and heading from gyroscopes and similar instruments and by providing appropriate control signals to the plane's steering mechanism. Automatic navigation systems and instrument landing systems operate by using radio signals from ground beacons that provide the aircraft with course directions for guidance. When an airplane is within the traffic pattern for ground control, its human pilot normally assumes control.

Examples of automated rail transportation include American urban mass-transit systems such as BART (Bay Area Rapid Transit) in San Francisco; MARTA (Metropolitan Atlanta Rapid Transit Authority) in Atlanta, Ga.; and the Metrorail in Washington, D.C. The BART system serves as a useful example; it consists of more than 75 miles (120 kilometres) of track, with about 100 trains operating at peak hours between roughly 30 stations. The trains sometimes attain speeds of 80 miles per hour with intervals between trains of as little as 90 seconds. In each train there is one operator whose role is that of an observer and communicator and who can override the automatic system in case of emergency. The automatic system protects the trains by assuring a safe distance between them and by controlling their speed. Another function of the system is to control train routings and make adjustments in the operation of each train to keep the entire system operating on schedule.

As a train enters the station, it automatically transmits its identification, destination, and length, thus lighting up a display board for passenger information and transmitting information to the control centres. Signals are automatically returned to the train to regulate its time in the station and its running time to the next station. At the beginning of the day, an ideal schedule is determined; as the day progresses, the performance of each train is compared with the schedule, and adjustments are made to each train's operation as required. The entire system is controlled by two identical computers, so that if one malfunctions, the other assumes complete control. In the event of a complete failure of the computer control system, the system reverts to manual control.

Service Industries

Automation of service industries includes an assortment of applications as diverse as the services themselves, which include health care, banking and other financial services, government, and retail trade.

In health care the use of automation in the form of computer systems has increased dramatically to improve services and relieve the burden on medical staffs. In hospitals computer terminals on each nursing care floor record data on patient status, medications administered, and other relevant information. Some of these systems are used to perform additional functions such as ordering drugs from the hospital pharmacy and calling for orderlies. The system provides an official record of the nursing care given to patients and is used by the nursing staff to give a report at shift-change time. The computer system is connected to the hospital's business office so that proper charges can be made to each patient's account for services rendered and medicines provided.

Robotics is likely to play role in future health care delivery systems. The work that is done in hospitals by nurses, orderlies, and similar staff personnel includes some tasks that are routine and repetitive. Duties that might be automated using robots include making beds, delivering linens, and moving supplies between locations in the hospital. Robots might even become involved in certain aspects of patient care such as transporting patients to services in the hospital, passing food trays, and similar functions in which it is not critical that a hospital staff member be present. Research is currently under way to develop robots that would be capable of providing assistance to paraplegics and other physically handicapped persons. These robots would respond to voice commands and would be able to interpret statements in natural language (e.g., everyday English) from patients requesting service.

Banking and financial institutions have embraced automation in their operations—principally through computer technology—to facilitate the processing of large volumes of documents and financial transactions. The sorting of checks is done by optical character-recognition systems utilizing the special alphanumeric characters at the bottom of checks. Bank balances are computed and recorded using computer systems installed by virtually all financial institutions. Major banks have established electronic banking systems, including automatic teller machines. Located in places convenient for their customers, these automatic tellers permit users to complete basic transactions without requiring the assistance of bank personnel.

The stock exchanges rely on computer-automated systems to report transactions by ticker tape or closed circuit television. Brokerage houses use a computerized record-keeping system to track their customers' accounts. Monthly statements indicating the status of each account are automatically prepared and mailed to customers. Account executives employ video monitors in their offices, backed by a massive database, to retrieve current information on each stock almost instantaneously while they discuss possible purchases with their clients. Stock certificates are typically issued with machine-readable identifications to facilitate record keeping in sales and exchanges.

Credit card transactions have also become highly automated. Restaurants, retailers, and other organizations are using systems that automatically check the validity of a credit card and the credit standing of the cardholder in a matter of seconds as the customer waits for the transaction to be finalized. Some credit card transactions trigger immediate transfer of funds equal to the amount of the sale from the cardholder's account into the merchant's account.

Many government services are automated by means of computers and computerized databases. The Internal Revenue Service (IRS) of the U.S. government must review and approve the tax returns of millions of taxpayers each year. The detailed checking of returns is a task that has traditionally been done by large staffs of professional auditors on a sampled basis. In 1985 the IRS began using a computerized system to automate the auditing procedure for the 1984 returns. This system is programmed to perform the complex tax calculations on each return being audited. As tax laws change, the system is reprogrammed to do the calculations for the year. The computerized auditing system has permitted a substantial increase in the work load of the IRS auditing department without a corresponding increase in staffing.

Retail trade has seen a number of changes in its operations as a result of automation. Selling merchandise has typically been a very labour-intensive activity, with sales associates needed to assist customers with their selections and then finalize transactions at the cash register. Each transaction depletes the store's inventory, so the item purchased must be identified for reorder. Much clerical effort is expended by the store when inventory is managed by strictly manual procedures. Computerized systems have been installed in most modern retail stores to speed sales transactions and automatically update inventory records as the stock of each item is reduced. The systems are based on the Universal Product Code (UPC), originally adopted by the grocery industry in 1973, which uses optical bar-code technology. A bar code is an identification symbol consisting of a series of wide and narrow bars attached to each product that can be scanned and recognized by a bar-code reader. At the cash registers, these readers quickly identify the items being purchased. As the sales associate scans across the symbol using a laser beam reader, the product is properly identified and its price is entered into the sales transaction. Simultaneously, a record of the sale is made in the inventory files so that the item can be reordered.

Consumer Products

Consumer products ranging from automobiles to small appliances have been automated for the benefit of the user. Microwave ovens, washing machines, dryers, refrigerators, video recorders, and other modern household appliances typically contain a microprocessor that works as the computer controller for the device. The consumer operates the appliance by programming the controller to perform the required functions, including timing (ovens, dryers), power levels (microwave ovens), input channels (video recorders), and other cycle options (washing machines). The programming of the device is done simply by pressing a series of buttons in the proper sequence, so the user does not think of the procedure as programming a computer.

The automobile is an example of a highly automated consumer product. The modern automobile is typically equipped with several microprocessors that operate a variety of functions, including engine control (fuel-air ratio, for example), the clock, the radio, and cruise control.

Automation in Power Production

Power providers constantly deal with demands to increase productivity and reduce costs. This translates into the need for administrators, engineers, operators, planners, field crews, and others to collect and act on decision-making information. Power system vendors are following a trend to make devices smarter so they can create and communicate this information. The term "power system" describes the collection of devices that make up the physical systems that generate, transmit, and distribute power. The term "instrumentation and control (I&C) system" refers to the collection of devices that monitor, control, and protect the power system. Power system automation refers to using I&C devices to perform automatic decision making and control of the power system.

Electric Utilities constantly face the problems of providing power at competitive prices. Due to accidents, lightning and natural incidents power failures occur resulting in interruption of power supply. Therefore power systems should be monitored and controlled effectively to take timely actions and ensure uninterrupted supply of power. Power system automation is defined as the automatically collecting information and taking decision with the help of Instrumentation and Control devices. These devices monitor, protect and control the power system. It includes processes involved in generation and distribution of power. Power system Automation depends on data communications, local intelligence along with supervisory monitoring and control. In earlier times plant processes were manually supervised .This needed the operator/supervisor to be at the plant all the time which was not possible. Automation has solved this problem and can be mechanical, economical, computerized or a combination.

Automation at the substation involves use of SCADA and data communication. Substation Automation requires the use of SCADA, data communication to control, monitor and protect the functions and equipment's. It plays an important role in the global economy by eliminating the need for manpower and reducing costs involved.

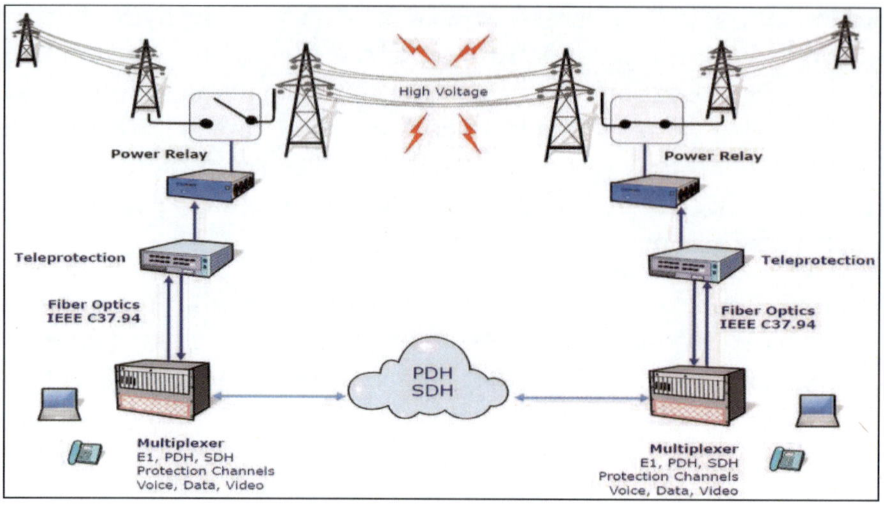

Power system Automation.

Types of Power System Automation

Substation Automation

At substations functions such as automatic reclosing, capacitor switching and reclosing are performed. Earlier these functions used to be performed with the help of devices such as relays, switches, lights and transducers. Currently Intelligent Electronic Devices [IED] consisting of programmable logic controllers and communication ports which send data and carry out control commands like meters, specialized sensor and relays. Inside the substation the expensive cables are replaced by local area networks to send data and execute control commands. Many techniques are available to send data outside the substation like telephone lines, dial up phone lines, cellular telemetry and fiber optic networks. This results in transmitting information at lesser cost per bit. PLC and other techniques enable to obtain more information about power system variables and equipment's. Information is obtained regarding location of fault and its analysis.

Distribution Automation

Distribution Automation systems are defined as those which allow to coordinate, analyse and operate components from further locations.

Features of Distribution Automation include:

- Improved public image,
- Better information,
- Increased System Efficiency,
- Consumer Satisfaction,

- Reduced Expenses,
- Goals of Distribution Automation:
  - Less Costs.
  - Improve service.
  - Improve reliability of Service.
  - Improve Government Relations.

Automation also includes:

- Energy Management systems: VAR, voltage and generation control and monitoring, load flow studies. Fault analysis and equipment maintenance.

Structure

Power system Automation consists of the following components:

- Data acquisition,
- Supervision,
- Electrical Protection.
- Control:
  - Control involves local and remote control. Local control is the steps that the controller can take such as synchronizing check and switching sequences. There is little risk of human error involved. Remote control controls the substation via SCADA master. Advantage offered is of faster operation. Examples include switching of circuit breaker in case of fault and request certain information via system. The operator has an overall view of the system enabling speedier decision making. Production losses are also reduced.
  - Electrical Protection: It is the most important function of automation to protect the devices and supervisors and to prevent damage in case of fault.
  - Measurement: Measurement involves electrical measurements such as power factor, current, voltage and harmonics. Analog measurements and disturbance recordings for fault analysis. This information about the station is obtained from the control room or central database. This prevents the operator/engineer from going to the station to obtain information thus assisting in worker safety and work is reduced. The information obtained can help in network studies, preventing faults and disturbances which cause large amount of losses.

- Monitoring: Sequence of events, status and condition monitoring. This can help in locating the fault and take preventive measures.
- Data Communication: Without data communication, power system automation cannot function. This can help to locate the fault and take preventive measures.

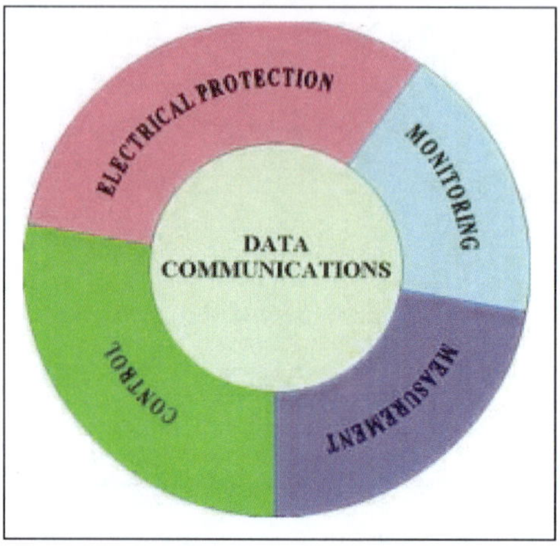

Functional structure of Power System Automation.

System Requirements

- An automation scheme shall have the following features.
- Controlling and monitoring the system equipment from a central point.
- Interface to SCADA system.
- Management of Energy.
- Control and monitoring of electrical equipment in a bay locally.

Power System Automation Architecture

Three divisions make up the structure of power automation system:

- Object Division: Object division of power system automation consists of RTU's, IED'S and relays. The current and potential transformers, transducers provide the analog inputs or digital inputs from IED'S, auxiliary contacts. They perform arithmetic and logical operations and provide outputs to SCADA master or back to switchgear. The component level consists of bay level (IED, RTU) and process level (information from current and potential transformer).
- Communication Network: It ensures that processed information, data and commands are transmitted error free and specially to the SCADA system, IED

and field instruments. The communication network is the most important part of power system automation and is the point of connection between SCADA system and bay level. The SCADA master might be in substation or central control room.

- SCADA Master: It performs the function of control of the entire automation system. It accepts data; takes decision regarding the data stores it and then sends commands to other remote devices. The SCADA master is made up of a PC and software package or a workstation. The installation maybe one master in control of several substations or the SCADA stations might form a LAN or WAN with the master controlling the sub-station on.

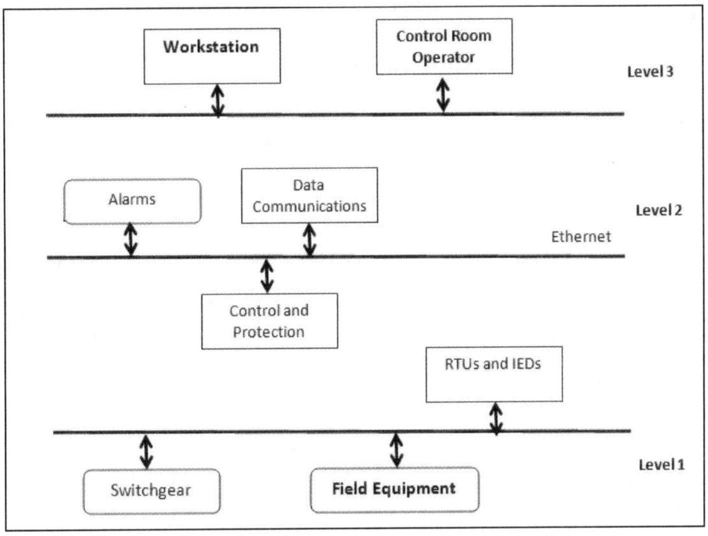

Architecture of Power system automation.

System Components

- I&C devices built using microprocessors are referred as intelligence electronic devices. Example of IED is a numerical relay. They are found in the substation and processes can be run into IED and communication handled through communication port on the computer.
- Bay module (Remote Terminal Unit) it is an IED installed at remote locations. It collects data (analog, digital and status) and transmits it in suitable form to SCADA master.
- Human Machine Interface includes a computer such as desktop computer or laptop with the necessary software installed in case of permanently installed HMI. A communication bus links the various devices.
- A link to SCADA system: It may be provided by interface unit, or part of HMI computer or IED. The link is used to communicate outside the substation.

Power System Integration

Power system Integration is defined as the act of transmitting data to and from the instrumentation and control devices and remote users. Substation integration is combining data from the IED to substation such that there is a single contact point for all data. IED's perform automation and integration in some cases. They perform data acquisition and IED control. The server plays the role of a client or master by collecting data and transmitting data to other devices as a slave in the communication industry. The data concentrator amasses data from various other devices. It transmits data between Intelligence electronic devices that are not directly connected. In this way through one data transfer data is transmitted. The archive client/server archives data obtained from devices and this archived data is retrieved in case of need.

Threats in Power System Automation

RTU, SCADA master, IED suffer from interruption of services and eavesdropping which are same problems faced by a normal computer.

Communication links again suffer from same problems faced by a computer such an interception of password and data. Information can be intercepted by unauthorized people. Similarly SCADA software suffers from attacks like intercepting, interruption and modification. The data must be encrypted to protect it from competition.

Counter Measures

The following measures are undertaken to eliminate threats:

- Equipment location: Since the equipment is located in remote locations they must be mounted or housed.
- Remote access: Password protection, encryption must be done for access to IED, RTU and relays.
- Human element: No unauthorized person must access the system without authentication or smart card access.

Advantages and Disadvantages of Automation

Advantages

- Better quality of service and reduction in labor required.
- Reduced operating costs.
- Flexible payment options and less outage minutes.
- Improved access to information for customers and to take decisions.
- Portable and less maintenance costs.

Disadvantages

- Absence of trained personnel,
- Large initial capital investments,
- Troubled Alarms.

Primary Applications

- Telemetry: It measures the amount of energy which is consumed. Data which is gathered daily from the bays is sent to the energy provider's server. The data is coded and sent through GPS, GSM making available at remote locations.
- Tele protection: Digital connection between two or three different facilities is ensured so that certain protection facilities can be carried out which are dependent on data from more than one substation.
- Equipment Supervision: This performs equipment maintenance functions such as direct voltage batteries and transformers. Performance improvement and wear and tear is prevented by change of settings.

Automated Retail

Automated Retail is the way of the future. Automated Retail is the use of a machine to sell retail products, but it is far more than this. Automated retail is a brand new sales channel for retailers and brands to reach customers in an innovative, exciting and non-traditional way.

Automated Retail has been growing exponentially within the last few years to an all-time high. Many retailers and national brands are looking for a way to build brand awareness and expand into untapped markets with minimal investment and low costs of entry.

We alluded to some of the benefits for rapidly scaling brands, but there are lots more:

Cut Costs

Mainly, retail automation allows you to reduce labor costs — more automation means fewer tasks done by hand, and a lesser need for the human capital to perform those tasks.

Retail robotics company Symbolic found that their technology successfully lowered labor costs by 80 percent — in a space up to 40 percent smaller, which equates to even more savings in terms of warehouse space optimization.

Not to mention, retail automation typically handles tedious, repetitive tasks. This will likely help improve job satisfaction, which could, in turn, reduce turnover rates and the expenses associated with filling those empty roles.

And despite what you might think, retail workers aren't necessarily afraid that robots will take their jobs. In fact, one Fair Workweek study found that 62 percent of employees think that these advanced technologies will actually create new job opportunities.

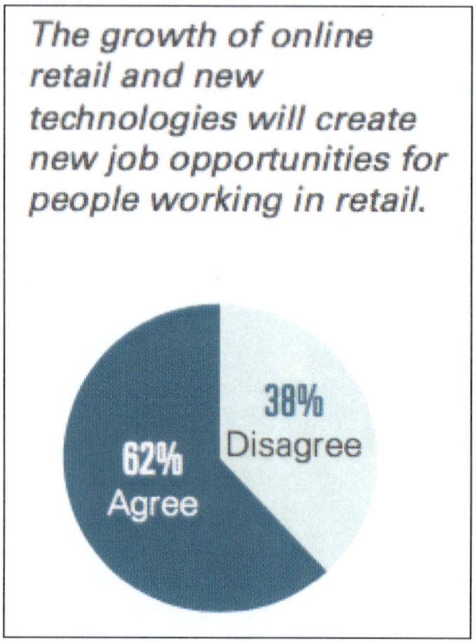

Become More Efficient

Symbiotic is just one example of how you can use retail automation to be more efficient in workforce management. But these efficiencies span across other areas of your retail business too.

Automation, in retail and across other industries, has one thing in common (when set up correctly): They save you time. This is because of a few reasons:

- Maintain around-the-clock, always-on business operations even if no one is physically working.
- Reduce the amount of time between tasks, since you're not waiting on a human to authorize the next step.
- Keep things moving. Even if someone calls out sick or is busy with another task, automation can keep operations running smoothly.
- Automation can limit the number of tasks needed to complete a workflow, reducing or eliminating steps altogether.

Improve the Customer Experience

Better efficiency means more chances to deliver timely customer support. And showing your customers you value their time is so important. According to Forrester, 77 percent of consumers say that valuing their time is the most important thing you can do to deliver a positive experience.

Automated customer support, personalized marketing automation, and quick order fulfillment are just a few ways this technology elevates the customer experience. And for high-growth brands, your customer base is constantly growing. Automation easily scales up to meet your needs, regardless of volume.

Plus, three-quarters of consumers want consistency across all channels. Retail automation can help you accomplish that by setting up those workflows or guidelines; this will ensure that customers have the same experience each time rather than a different one depending on the support specialist they're dealing with.

Drive Sales

More customer satisfaction typically leads to more sales. And isn't that the goal of every high-growth brand? Retail automation helps drive sales in a more proactive way as well.

Let's say you've set up an automated email sequence to send to customers who are interested in a specific type of product on your site. These personalized offers and promotions are likely to be far more effective than a generic one sent to your entire list. Consider this: 64 percent of consumers want personalized offers from retail brands.

Fuel Growth

To piggyback off of those increased sales, retail automation is also primed for attaining sustainable scale. As you've probably learned by now, sticking to manual processes isn't going to get your brand to the next growth stage. In fact, this old-school approach could even set your business back.

Automating workflows across your business will enable you to focus on optimizing processes for growth instead of having to build new ones. Digitally native brands need to be creative in their approach, and this means flexible workflows and automation that allow for fluctuations and outside-the-box thinking.

Safeguard your Biz

One major way retail automation can protect your business is by proactively detecting fraud. You can tell your software to send you an alert or require further identity verification for orders deemed high-risk. You can establish high-risk orders based on order value, browser location, or shipping location, among others.

This is huge: Fraud is the top concern for retailers, according to the National Retail Federation. And it's on the rise – online fraud in particular. And another study found that 70 percent have machine-learning analytics very high on the priority list for fraud mitigation investments.

Reduce Human Error

Less human error means cleaner, more accurate data (not to mention fewer mistakes in day-to-day operations). Yet 62 percent of manufacturers still use pen and paper to track their processes, according to one survey.

Moving to automated processes can improve your data quality and mitigate errors.

Rapidly scaling brands rely on data intelligence. Being able to glean insights from your business metrics, and then implement strategies to capitalize on growth opportunities and identify issues, is critical. In-store, robots scan the aisles and send notifications when discrepancies arise.

Automation also means data is readily available. Rather than having to manually update spreadsheets before you can pull reports, automation keeps data up-to-date in real time. You can even automate the retail reporting process. This'll help keep everyone on the same page, by distributing the same data set across teams internally.

Examples of Retail Automation

When it comes to eCommerce and retail, automation happens in two main ways:

- Behind-the-scenes: These are internal and don't directly apply to the customer.
- In-store: For the purposes of this discussion, we'll consider in-store anything that happens in a brick-and-mortar storefront or on your eCommerce site.

Behind-the-scenes Retail Automation

Many business operations and processes can be automated. These are what keep your business running smoothly, so you and your colleagues can focus on new growth opportunities.

A few key areas for behind-the-scenes automation for retailers:

- Warehouse management: You've likely heard of robots in retail and how useful they are in warehouse management. Amazon has been using them for order fulfillment, and Walmart has FAST Unloaders which optimize warehouse space. Robots maximize warehouse space by allowing you to reach new places and organize products in such a way that humans wouldn't be able to manage. Plus, they're more efficient than humans when it comes to pick, pack, and ship. In addition to warehouse robots, you can build automation that fulfill and route orders based on product availability and customer location. This ensures the fastest and most affordable shipping options — well for you and your customers.

- Inventory management: Closely related to warehouse management, automation work on the inventory control side of things in a few ways, including automated purchasing systems, changes in product demand, and low stock alerts. Automated product reporting is also especially valuable for high-growth brands: This helps you determine top sellers and revenue opportunities to build upon your growth.
- Marketing: We mentioned this before, but consumers are demanding more personalized interactions and promotions from brands. You can build out automations to tag customers with specific attributes, and automatically trigger marketing campaigns based on actions they take. For example, if someone hasn't made a purchase in 6 months, you can create an automation to remind them that it's time to go shopping again.
- Accounting: You can automate financial processes both on the accounts payable and accounts receivable side of things. For cash inflow, automated billing and payment processing keeps money coming in and prevents customers from waiting. When it comes to outflow, you can automate invoice and purchase order payments to avoid delays in your supply chain.
- Multi-channel management: If you're like many high-growth brands, you're selling on more than one channel. This could be your online store, wholesale, pop-up shops, third-party marketplaces, or even social selling. Multi-channel inventory management is critical. If stock levels aren't automatically updated across channels, for instance, you run the risk of stockouts and falling short of customer expectations.

In-store Retail Automation

"Customer-facing" automation is also invaluable to your business. And while we've discussed robots on the back-end of things, robots in the retail industry can also be effective in supporting scalability in the in-store environment.

One way is by providing a positive and efficient customer experience. Lowe's introduced an in-store robot, LoweBot, to help customers with wayfinding and product information.

Robots in stores aren't the only way to implement retail store automation. The self-serve trend is also major. We're talking things like self-checkout, kiosks, and vending machines. One survey found that two-thirds of consumers want self-serve options — and most believe it will improve the shopping experience.

Some examples of self-serve that come to mind are Redbox kiosks or the consumer electronics vending machines becoming airport staples. But forward-thinking brands are taking more innovative approaches.

Amazon Go is one well-known and completely automated retail store, Greenfire popped up in Seattle with the same cashless premise (prior to Amazon's debut), and Carvana has turned car-buying into a trip to the vending machine. Walmart is even introducing self-serve "lockers" where customers can pick up their online orders.

When it comes to your eCommerce store, chatbots are automated messaging tools that can respond to customer queries before directing them to the appropriate human. You might be able to answer common queries without ever getting an actual person involved.

How Retail Automation Supports Sustainable Growth?

Growing your business doesn't require a huge investment in human capital. With retail automation, you can increase the productivity and quality of work your existing team is capable of — and empower them to make smarter, more confident decisions to support scale into the future.

Your business is only as successful as the team behind it. Give them the freedom to think strategically, explore new ideas, and build memorable customer experiences that will drive growth for years to come.

Automation in Mining

Automation is the new hot topic in the mining industry. A recent poll conducted by Mining IQ revealed that 77% of mining specialists think automation is a high priority, and 40% of these specialists say that automation is "now more important than ever." A Google search will turn up dozens of current conferences on the topic, an indication of the pace at which mining companies are engaging in automated and robotic technology. But the subject of mine site automation stirs up controversy as well as excitement.

While mining companies are seeking mining automation technologies for the operational and financial efficiencies they offer, employees fear being made obsolete by innovations such as driverless trucks and trains. Remote monitoring and control of equipment allows miners to automate industrial processes like blasting, drilling, and transportation.

Efficiencies are gained through continuous, consistent operations, improved communications, and reduced infrastructure. Mining companies claim automation isn't really eliminating jobs but rather creating new, different ones. Supporters maintain that while traditional entry level and field positions may be lost, personnel can be shifted into new roles monitoring and controlling the automated processes. If a driverless truck stops, for example, site managers are needed to coordinate mine communications to report the stoppage and dispatch an operator to check and manually restart the vehicle. Proponents also point out that mine site safety can be improved by removing personnel from dangerous environments and placing them in remote control rooms where they can operate equipment from a safe distance.

A new report issued by the Australian Workforce and Productivity Agency (AWPA) suggests workers with specialized skills in remotely controlled and automated systems will find opportunities as the trend toward automation increases. For current employees, this means a trend away from lower-skilled jobs into more highly skilled technical jobs and therefore retraining and re-education. Other potential opportunities may arise in the areas of lab automation, information management, and data analysis. The need for rapid data capture and accurate analysis of sample data is one of the biggest challenges in mining. Timely sample analyses are critical to determine ore boundaries and drilling targets in order to focus on the highest quality deposits.

Modern mining companies looking to improve operational efficiency rely on sophisticated laboratory information management systems (LIMS) for quality control, increased productivity, data management, and compliance with product and environmental safety standards. Lab managers and technicians will be needed as LIMS expand to small and mid-tier companies trying to stay competitive. Employees skilled in data analysis can develop predictive models that help mine operations become proactive rather than reactive, a significant advantage when substantial investments are at stake.

Reasons to Automate Mines

- Advanced software automation creates better predictability. It's estimated that less than one percent of all data collected and generated by existing mining exploratory technology is not used. This is mostly due to the inability to process and present data in real time. Software automation can make this wealth of data more useful to investors, managers, and stakeholders by providing more exploratory information.
- Software automation can make adjustments much easier. Mines are by their nature unpredictable. Conditions change over the course of a single mine's operation. Intelligent software automation takes the guesswork out of adjustments, avoiding production slowdowns and mitigating costs.
- Software and robotic automation cut costs: More efficient extraction of resources from mines and the better use of information from data collection devices make it easier to cut costs and learn how to create a leaner, more efficient mining operation overall.

- Robotic solutions improve safety: Workers in mines can utilize robotic, automated equipment better and more efficiently, creating safer conditions in mines. Some robotic technology, such as tele-operated mining equipment, can be used in hazardous areas that humans aren't suited for.
- Better fuel efficiency: The elimination of guesswork in the proper use of mining equipment and heavy machinery improves fuel efficiency, which can cut down on operating costs significantly.
- Less downtime due to maintenance: Both the use of automated software to view complex data in real time and the use of robotic technology in mines cuts down on unscheduled maintenance and reduce the downtime that maintenance work inevitably causes.
- Overall reduction of overhead costs: We've talked about cutting costs, but automation of some functions of mine operation can mitigate the losses that labor shortages sometimes cause, especially in an age of lower commodity prices. Some simple jobs can now be done by robots instead of workers.
- Improved exploration processes: Before the mining operation starts in earnest, software automation and robotic technology can help executives understand what exactly is under the surface of the earth before investing in the operation.
- Improved communication: Communication between stakeholders, executives, and the workers themselves lead to more efficient, better connected mines as a whole.
- Reach remote mining locations. Sometimes getting equipment and miners to a site is the most difficult job in a mining operation. Using automated equipment, stakeholders can make better decisions on to extract resources from remote mining locations.

Mine Automation Software

In order to gain more control over their operations, mining companies may implement mining automation software or processes. Reports generated by mine automation software allow administrators to identify productivity bottlenecks, increase accountability, and better understand return on investment.

Mining Equipment Automation

Addressing concerns about how to improve productivity and safety in the mine site, some mine companies are turning to equipment automation consisting of robotic hardware and software technologies that convert vehicles or equipment into autonomous mining units.

Mine equipment automation comes in four different forms: remote control, teleoperation, driver assist, and full automation.

Remote Control

Remote control mining equipment usually refers to mining vehicles such as excavators or bulldozers that are controlled with a handheld remote control. An operator stands in line-of-sight and uses the remote control to perform the normal vehicle functions. Because visibility and feel of the machine are heavily reduced, vehicle productivity is generally reduced as well using remote control. Remote control technology is generally used to enable mining equipment to operate in dangerous conditions such as unstable terrain, blast areas or in high risk areas of falling debris, or underground mining. Remote control technology is generally the least expensive way to automate mining equipment making it an ideal entry point for companies looking to test the viability of robotic technology in their mine.

Teleoperated Mining Equipment

Teleoperated mining equipment refers to mining vehicles that are controlled by an operator at a remote location with the use of cameras, sensors, and possibly additional positioning software. Teleoperation allows an operator to further remove themselves from the mining location and control a vehicle from a more protected environment. Joysticks or other handheld controls are still used to control the vehicle's functions, and operators have greater access to vehicle telemetry and positioning data through the teleoperation software. With the operator removed from the cab, teleoperated mining vehicles may also experience reduced productivity; however, the operator has a better vantage point than remote control from on-vehicle cameras and sensors and is further removed from potentially dangerous conditions.

Driver Assist

"Driver assist" refers to partly automated control of mining machines. Only some of the functions are automated and operator intervention is needed. Common functions include both spotting assist and collision avoidance systems.

Full Automation

"Full automation" can refer to the autonomous control of one or more mining vehicles. Robotic components manage all critical vehicle functions including ignition, steering, transmission, acceleration, braking, and implement control (i.e. blade control, dump bed control, excavator bucket and boom, etc.) without the need for operator intervention. Fully autonomous mining systems experience the most productivity gains as software controls one or more mining vehicles allowing operators to take on the role of mining facilitators, troubleshooting errors and monitoring efficiency.

Benefits

The benefits of mining equipment automation technologies are varied but may include: improved safety, better fuel efficiency, increased productivity, reduced unscheduled maintenance, improved working conditions, better vehicle utilization, and reduced driver fatigue and attrition. Automation technologies are an efficient way to mitigate the effects of widespread labor shortages for positions such as haul truck driver. In the face of falling commodity prices, many mining companies are looking for ways to dramatically reduce overhead costs while still maintaining site safety and integrity; automation may be the answer.

Drawbacks

Critics of vehicle automation often focus on the potential for robotic technology to eliminate jobs while proponents counter that while some jobs will become obsolete (normally the dirty, dangerous, or monotonous jobs), others will be created. Communities supporting underprivileged workers that rely on entry level mining positions are worried about and are calling for social responsibility as mining companies transition to automation technologies that promise to increase productivity in the face of falling commodity prices. Risk averse mining companies are also reluctant to commit large amounts of capital to an unproven technology, preferring more often to enter the automation scene at lower, more inexpensive levels such as remote control.

Examples of Autonomous Mining Equipment

Mine of the Future

Rio Tinto Group embarked on their Mine of the Future initiative in 2008. From a control center in Perth, Rio Tinto employees operate autonomous mining equipment in Australia's remote but mineral rich Pilbara region. The autonomous mining vehicles reduce the footprint of the mining giant while improving productivity and vehicle utilization. As of June 2014, Rio Tinto's autonomous mining fleet reached the milestone of 200 million tons hauled. Rio Tinto also operate a number of autonomous blast hole drill rigs.

Bingham Canyon Mine

Located near Salt Lake City, Utah, the Bingham Canyon Mine (Kennecott Utah Copper/Rio Tinto) is one of the largest open pit mine in the world and one of the world's largest copper producers. In April 2013, the mine experienced a catastrophic landslide that halted much of the mine's operations. As part of the cleanup efforts and to improve safety, mine administrators turned to remote control excavator, dozers and teleremote blast hole drills to perform work on the highly unstable terrain areas. Robotic technology helped Kennecott to reduce the steeper, more dangerous areas of the slide to allow manned vehicles access for cleanup efforts.

Automation of Underground Works in China

German company "EEP Elektro-Elektronik Pranjic" delivered and put into operation more than 60 sets of advanced automatic control for underground coal mining for the period ~ 2006-2016. For the first time completely deserted coal mining technology has been used by the Chinese concern "China National Coal Group Corp. (CME)" at the mine "Tang Shan Gou" (longwall mining, shearers, three lava, depth 200 m), and at the mine "Nan Liang" (one plow, depth 100 m). Both coal mines have coal layer thickness 1-1.7 m. Monitoring the harvesting is carried out by means of video cameras (in real time with signal transmission over optical fiber). Typically, an underground staff is required to monitor the production process and for carrying out repairs. Automation has improved the safety and economic performance.

Automation in Car

A self-driving car (sometimes called an autonomous car or driverless car) is a vehicle that uses a combination of sensors, cameras, radar and artificial intelligence (AI) to travel between destinations without a human operator. To qualify as fully autonomous, a vehicle must be able to navigate without human intervention to a predetermined destination over roads that have not been adapted for its use.

Companies developing and testing autonomous cars include Audi, BMW, Ford, Google, General Motors, Tesla, Volkswagen and Volvo. Google's test involved a fleet of self-driving cars -- including Toyota Prii and an Audi TT -- navigating over 140,000 miles of California streets and highways.

Working of Self-driving Cars

AI technologies power self-driving car systems. Developers of self-driving cars use vast amounts of data from image recognition systems, along with machine learning and neural networks, to build systems that can drive autonomously.

The neural networks identify patterns in the data, which is fed to the machine learning algorithms. That data includes images from cameras on self-driving cars from which the neural network learns to identify traffic lights, trees, curbs, pedestrians, street signs and other parts of any given driving environment.

For example, Google's self-driving car project, called Waymo, uses a mix of sensors, Lidar (light detection and ranging -- a technology similar to radar) and cameras and combines all of the data those systems generate to identify everything around the vehicle and predict what those objects might do next. This happens in fractions of a second. Maturity is important for these systems. The more the system drives, the more data it can incorporate into its deep learning algorithms, enabling it to make more nuanced driving choices.

The following outlines how Google Waymo vehicles work:

- The driver (or passenger) sets a destination. The car's software calculates a route.
- A rotating, roof-mounted Lidar sensor monitors a 60-meter range around the car and creates a dynamic three-dimensional (3D) map of the car's current environment.
- A sensor on the left rear wheel monitors sideways movement to detect the car's position relative to the 3D map.
- Radar systems in the front and rear bumpers calculate distances to obstacles.
- AI software in the car is connected to all the sensors and collects input from Google Street View and video cameras inside the car.
- The AI simulates human perceptual and decision-making processes using deep learning and controls actions in driver control systems, such as steering and brakes.
- The car's software consults Google Maps for advance notice of things like landmarks, traffic signs and lights.
- An override function is available to enable a human to take control of the vehicle.

Cars with Self-driving Features

Google's Waymo project is an example of a self-driving car that is almost entirely autonomous. It still requires a human driver to be present but only to override the system when necessary. It is not self-driving in the purest sense, but it can drive itself in ideal conditions. It has a high level of autonomy. Many of the cars available to consumers today have a lower level of autonomy but still have some self-driving features. The self-driving features that are available in many production cars as of 2019 include the following:

- Hands-free steering centers the car without the driver's hands on the wheel. The driver is still required to pay attention.
- Adaptive cruise control (ACC) down to a stop automatically maintains a selectable distance between the driver's car and the car in front.
- Lane-centering steering intervenes when the driver crosses lane markings by automatically nudging the vehicle toward the opposite lane marking.

Levels of Autonomy in Self-driving Cars

The U.S. National Highway Traffic Safety Administration (NHTSA) lays out six levels of automation, beginning with Level 0, where humans do the driving, through driver assistance technologies up to fully autonomous cars. Here are the five levels that follow Level 0 automation:

- Level 1: An advanced driver assistance system (ADAS) aid the human driver with steering, braking or accelerating, though not simultaneously. An ADAS includes rearview cameras and features like a vibrating seat warning to alert drivers when they drift out of the traveling lane.
- Level 2: An ADAS that can steer and either brake or accelerate simultaneously while the driver remains fully aware behind the wheel and continues to act as the driver.
- Level 3: An automated driving system (ADS) can perform all driving tasks under certain circumstances, such as parking the car. In these circumstances, the human driver must be ready to retake control and is still required to be the main driver of the vehicle.
- Level 4: An ADS can perform all driving tasks and monitor the driving environment in certain circumstances. In those circumstances, the ADS is reliable enough that the human driver needn't pay attention.
- Level 5: The vehicle's ADS acts as a virtual chauffeur and does all the driving in all circumstances. The human occupants are passengers and are never expected to drive the vehicle.

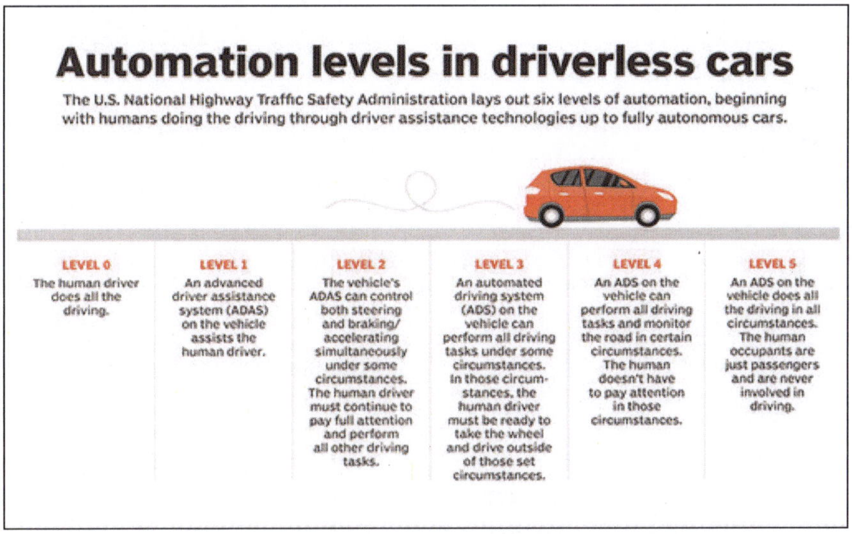

Automation spans from zero automation to fully autonomous.

Uses

As of 2019, carmakers have reached Level 4. Manufacturers must clear a variety of technological milestones, and several important issues must be addressed before fully autonomous vehicles can be purchased and used on public roads in the United States.

Even though cars with Level 4 autonomy aren't available for public consumption, they are in used in other ways.

For example, Google's Waymo partnered with Lyft to offer a fully autonomous commercial ride-sharing service called Waymo One. Riders can hail a self-driving car to bring them to their destination and provide feedback to Waymo. The cars still include a safety driver in case the ADS need to be overridden. The service is only available in the Metro Phoenix area as of late 2019 but is looking to expand to cities in Florida and California.

Autonomous street-sweeping vehicles are also being produced in China's Hunan province, meeting the Level 4 requirements for independently navigating a familiar environment with limited novel situations.

Projections from manufacturers vary on when Level 4 and 5 vehicles will be widely available. Ford and Volvo both project a 2021 release of a Level 4 vehicle for public consumption. Tesla CEO Elon Musk, being a pioneer of both self-driving and electric cars, has claimed that his company will have Level 5 vehicles ready as early as 2020. A successful Level 5 car must be able to react to novel driving situations as well or better than a human can.

The Pros and Cons of Self-driving Cars

The top benefit touted by autonomous vehicle proponents is safety. A U.S. Department of Transportation (DOT) and NHTSA statistical projection of traffic fatalities for 2017 estimated that 37,150 people died in motor vehicle traffic accidents that year. NHTSA estimated that 94% of serious crashes are due to human error or poor choices, such as drunk or distracted driving. Autonomous cars remove those risk factors from the equation -- though self-driving cars are still vulnerable to other factors, such as mechanical issues, that cause crashes.

If autonomous cars can significantly reduce the number of crashes, the economic benefits could be enormous. Injuries impact economic activity, including $57.6 billion in lost workplace productivity and $594 billion due to loss of life and decreased quality of life due to injuries, according to NHTSA.

In theory, if the roads were mostly occupied by autonomous cars, traffic would flow smoothly, and there would be less traffic congestion. In fully automated cars, the occupants could do productive activities while commuting to work. People who aren't able to drive due to physical limitations could find new independence through autonomous vehicles and would have the opportunity to work in fields that require driving.

Autonomous trucks have been tested in the U.S. and Europe to enable drivers to use autopilot over long distances, freeing the driver to rest or complete tasks and improving driver safety and fuel efficiency. This initiative, called truck platooning, is powered by

ACC, collision avoidance systems and vehicle-to-vehicle communications for cooperative ACC (CACC).

The downsides of self-driving technology could be that riding in a vehicle without a driver behind the steering wheel may be unnerving -- at least at first. But, as self-driving capabilities become commonplace, human drivers may become overly reliant on the autopilot technology and leave their safety in the hands of automation, even when they should act as backup drivers in case of software failures or mechanical issues.

In one example from March 2018, Tesla's Model X sport utility vehicle (SUV) was on autopilot when it crashed into a highway lane divider. The driver's hands were not on the wheel, despite visual warnings and an audible warning to put his hands back on the steering wheel, according to the company. Another crash occurred when a Tesla's AI mistook the side of a truck's shiny reflection for the sky.

Self-driving Car Safety and Challenges

Autonomous cars must learn to identify countless objects in the vehicle's path, from branches and litter to animals and people. Other challenges on the road are tunnels that interfere with the Global Positioning System (GPS), construction projects that cause lane changes or complex decisions, like where to stop to allow emergency vehicles to pass.

The systems need to make instantaneous decisions on when to slow down, swerve or continue acceleration normally. This is a continuing challenge for developers, and there are reports of self-driving cars hesitating and swerving unnecessarily when objects are detected in or near the roadways.

This problem was evident in a fatal accident in March 2018, which involved an autonomous car operated by Uber. The company reported that the vehicle's software identified a pedestrian but deemed it a false positive and failed to swerve to avoid hitting her. This crash caused Toyota to temporarily cease its testing of self-driving cars on public roads, but its testing will continue elsewhere. The Toyota Research Institute is constructing a test facility on a 60-acre site in Michigan to further develop automated vehicle technology.

With crashes also comes the question of liability, and lawmakers have yet to define who is liable when an autonomous car is involved in an accident. There are also serious concerns that the software used to operate autonomous vehicles can be hacked, and automotive companies are working to address cybersecurity risks.

Carmakers are subject to Federal Motor Vehicle Safety Standards (FMVSS), and NHTSA reported that more work must be done for vehicles to meet those standards.

In China, carmakers and regulators are adopting a different strategy to meet standards and make self-driving cars an everyday reality. The Chinese government is beginning to redesign urban landscapes, policy and infrastructure to make the environment more self-driving car-friendly. This includes writing rules about how humans move around and recruiting mobile network operators to take on a portion of the processing required giving self-driving vehicles the data they need to navigate. "National Test Roads" would be implemented. The autocratic nature of the Chinese government makes this possible, which bypasses the litigious democracy that tests are funneled through in America.

The path toward self-driving cars began with incremental automation features for safety and convenience before the year 2000, with cruise control and antilock brakes. After the turn of the millennium, advanced safety features, including electronic stability control, blind-spot detection, and collision and lane shift warnings, became available in vehicles. Between 2010 and 2016, advanced driver assistance capabilities, such as rearview video cameras, automatic emergency brakes and lane-centering assistance, emerged, according to NHTSA.

Since 2016, self-driving cars have moved toward partial autonomy, with features that help drivers stay in their lane, along with ACC technology and the ability to self-park.

Fully automated vehicles are not publicly available yet and may not be for many years. In the U.S., NHTSA provides federal guidance for introducing ADSes onto public roads and, as autonomous car technologies advance, so will the department's guidance.

Self-driving cars are not yet legal on most roads. In June 2011, Nevada became the first jurisdiction in the world to allow driverless cars to be tested on public roadways; California, Florida, Ohio and Washington, D.C., have followed in the years since.

The history of driverless cars goes back much further than that. Leonardo da Vinci designed the first prototype around 1478. Da Vinci's car was designed as a self-propelled robot powered by springs, with programmable steering and the ability to run preset courses.

References

- Computer-integrated-manufacturing, technology-automation- 24857: britannica.com, Retrieved 07 April, 2019
- What-is-automated-retail, how-it-works: innovativevendingsolutions.com, Retrieved 10 May, 2019
- Retail-automation-to-fuel-growth, learning-center: stitchlabs.com, Retrieved 19 June, 2019
- Is-automation-the-future-of-mining, mining: thermofisher.com, Retrieved 16 March, 2019
- 10-reasons-automate-mine: micromine.com, Retrieved 25 August, 2019
- Automated-highway-systems, Civil-Engineering: seminarsonly.com, Retrieved 14 June, 2019
- Top-10-benefits-of-automating-your-home: freshome.com, Retrieved 25 January, 2019
- 5-technologies-that-would-change-waste-management-industry- 22289: roboticsandautomationnews.com, Retrieved 05 July, 2019

Permissions

All chapters in this book are published with permission under the Creative Commons Attribution Share Alike License or equivalent. Every chapter published in this book has been scrutinized by our experts. Their significance has been extensively debated. The topics covered herein carry significant information for a comprehensive understanding. They may even be implemented as practical applications or may be referred to as a beginning point for further studies.

We would like to thank the editorial team for lending their expertise to make the book truly unique. They have played a crucial role in the development of this book. Without their invaluable contributions this book wouldn't have been possible. They have made vital efforts to compile up to date information on the varied aspects of this subject to make this book a valuable addition to the collection of many professionals and students.

This book was conceptualized with the vision of imparting up-to-date and integrated information in this field. To ensure the same, a matchless editorial board was set up. Every individual on the board went through rigorous rounds of assessment to prove their worth. After which they invested a large part of their time researching and compiling the most relevant data for our readers.

The editorial board has been involved in producing this book since its inception. They have spent rigorous hours researching and exploring the diverse topics which have resulted in the successful publishing of this book. They have passed on their knowledge of decades through this book. To expedite this challenging task, the publisher supported the team at every step. A small team of assistant editors was also appointed to further simplify the editing procedure and attain best results for the readers.

Apart from the editorial board, the designing team has also invested a significant amount of their time in understanding the subject and creating the most relevant covers. They scrutinized every image to scout for the most suitable representation of the subject and create an appropriate cover for the book.

The publishing team has been an ardent support to the editorial, designing and production team. Their endless efforts to recruit the best for this project, has resulted in the accomplishment of this book. They are a veteran in the field of academics and their pool of knowledge is as vast as their experience in printing. Their expertise and guidance has proved useful at every step. Their uncompromising quality standards have made this book an exceptional effort. Their encouragement from time to time has been an inspiration for everyone.

The publisher and the editorial board hope that this book will prove to be a valuable piece of knowledge for students, practitioners and scholars across the globe.

Index